HIGH PRESSURE
LIQUID CHROMATOGRAPHY

Biochemical and Biomedical
Applications

HIGH PRESSURE
LIQUID
CHROMATOGRAPHY

Biochemical and Biomedical
Applications

PHYLLIS R. BROWN

Department of Biochemical Pharmacology
Division of Biological and Medical Sciences
Brown University
Providence, Rhode Island

ACADEMIC PRESS   1973   *New York and London*

A Subsidiary of Harcourt Brace Jovanovich, Publishers

ACADEMIC PRESS, INC.
111 Fifth Avenue, New York, New York 10003

United Kingdom Edition published by
ACADEMIC PRESS, INC. (LONDON) LTD.
24/28 Oval Road, London NW1

LIBRARY OF CONGRESS CATALOG CARD NUMBER: 72-77361

PRINTED IN THE UNITED STATES OF AMERICA

*To my husband, without whose infinite patience,
continuous encouragement, and constant support this
book could never have been written*

CONTENTS

PREFACE

In the past few years, the development of commercially available high pressure liquid chromatography systems has opened up new horizons for the analysis of nonvolatile, reactive compounds. Although gas chromatography has proved highly sucessful in organic chemistry, its applications are limited. Many of the thermally labile, complex molecules, which are difficult to analyze by gas chromatography, are important in biochemistry and the biomedical field. Other techniques, such as thin-layer, paper, and column chromatography, while useful, have never been completely satisfactory. High pressure liquid chromatography overcomes many of these difficulties and provides excellent resolution with high speed and sensitivity for separating high molecular weight, polar compounds. Moreover, its applications are not limited to this type of sample; separations of all compounds that are soluble in a liquid phase can be obtained. This technique can be useful not only in biochemistry but also in inorganic, organic, and clinical chemistry.

The purpose of this book is to present simply and concisely basic information on high pressure liquid chromatography. It should serve as an introduction for those unfamiliar with this technique as well as a reference book for those who are currently using it. I sincerely believe that high pressure liquid chromatography will prove to be an invaluable aid in all biochemical and biomedical research and hope that this book will encourage its use.

Since no analytical method is perfect, the pitfalls likely to be encountered are discussed to help newcomers avoid many of the problems experienced by pioneers in the field. Because my experience has been mainly in the use of anion-exchange, high pressure liquid chromatography to separate nucleotides, most of the examples given are from this area.

Much to my surprise, reviewing the literature I found that the great majority of articles have been concerned with the development of the instrumentation and that comparatively little work has been published on the applications of this technique to biochemical and biomedical research. In the next few years however, when researchers in chemical, biochemical, and medical disciplines fully realize the great potential of high pressure liquid chromatography, there should be a literature explosion such as was experienced in the field of gas chromatography in the 1960's.

I wish to thank the many people who graciously contributed information, chromatograms, and assistance in the preparation of this book. I would like particularly to thank Dr. R. E. Parks, Jr., who appreciated the need for instrumentation to measure nucleotide pools and anticipated the value of high pressure liquid chromatography for this type of investigation. Without his constant encouragement and support, this work would not have been possible. I would also like to thank Drs. John O. Edwards and Ralph Miech for their many helpful discussions; Drs. Andrew Clifford, Eric Scholar, J. Arly Nelson, James Little, Richard Henry, Csaba Horvath, and Herman Shmukler for their gracious cooperation and for making available to me data prior to publication; Dr. Dennis Gere for his expert technical assistance and unstinting help when needed; and Miss Elise Bideaux and Miss Marijo Newman for laboratory assiatance. To Mr. Jonathan Gell are extended my special thanks for his technical skills which kept our chromatograph in excellent condition, enabling its continuous operation during the preparation of this book. I am grateful to Mrs. Mary Bonaventure for her fine typing job and to Dr. Marie Sauer and the staff of the drafting office of the engineering department for their excellent work in preparing the illustrations. I am indebted to Drs. Edwards, Horvath, Little, Clifford, Henry, and Gere for taking time from their busy schedules to read this manuscript and for giving me the benefit of their knowledge and experience. Their many helpful comments, suggestions, and criticisms were gratefully appreciated. I am also indebted to U.S.P.H.S. Grant 16538-01 for support of research in this field.

Finally, I wish to thank my family for their patience, consideration, and cooperation. Without their understanding, it would have been impossible to complete this task.

Phyllis R. Brown

ABBREVIATIONS

The following abbreviations are used: AMP, ADP, ATP = adenosine 5'-mono-, 5'-di-, and 5'-triphosphate; cAMP = cyclic 3',5'-adenosine monophosphate; GMP, GDP, GTP = 5'-phosphates of guanosine; cGMP = cyclic 2',3'-guanosine monophosphate; UMP, UDP, UTP = 5'-phosphates of uridine; CMP, CDP, CTP = 5'-phosphates of cytidine; TMP, TDP, TTP = 5'-phosphates of deoxythymidine; 6 MMPRMP = 5'-monophosphate of 6-methyl mercaptopurine riboside; TGMP = 5'-monophosphate of 6-thioguanosine; XMP = 5'-monophosphate of xanthosine; IMP, IDP = 5'-phosphates of inosine; UDPG = uridine diphosphoglucose; UDPGA = uridine 5'-diphosphoglucuronic acid; NAD, NADH = the oxidized and reduced forms of nicotinamide-adenine dinucleotide; TCA = trichloroacetic acid; PCA = perchloric acid.

CHAPTER 1

INTRODUCTION

I. Description

In the literature, liquid chromatographic systems which feature high inlet pressures and high sensitivity are variously described as high pressure, high speed, high performance, high efficiency, or high sensitivity liquid chromatography. In all cases, the reference is to liquid chromatographic systems which utilize high pressure pumps and sensitive, low dead-volume detectors. No one title is completely descriptive. In fact, a better description of this type of separation is liquid chromatography that affords high resolution, high speed, high efficiency, and high sensitivity at the same time. Although the various terms will be used interchangeably in this book, I have chosen the title "High Pressure Liquid Chromatography" for two reasons. First, it was the term I was introduced to originally and have used continually for the past three years. It has become an indispensable part of my vocabulary. Second, the use of pressure to pump the eluent through the column is the main feature that distinguishes this particular technique from other types of chromatography. In other modes of modern liquid chromatography high performance or high efficiency at high speed may be achieved, but for this method high inlet pressures are a necessity. Many people, when speaking of liquid chromatographic systems, are referring only to ion exchange chromatography such as is used in the amino acid analyzer or the nucleic acid analyzer. However, here the term liquid chromatography will be used in the

1

broader sense and will include liquid–liquid (partition), liquid–solid (adsorption), and gel permeation as well as ion exchange chromatography.

All types of chromatography are based on the phenomenon that each component in a mixture interacts differently with its environment from other components under the same conditions. Basically, chromatography is a separation technique, a differential migration process where the sample components are selectively retained by a stationary phase. It involves separation due to differences in the equilibrium distribution of sample components between two immiscible phases: a moving or mobile phase and a stationary phase. The sample components must be in the mobile phase to migrate through the chromatographic system. Since the velocity of migration is a function of equilibrium distribution, the components which favor the stationary phase migrate slower than those whose distributions favor the mobile phase. Therefore, separation of components is caused by different velocities of migration as a result of differences in equilibrium distributions.

Chromatographic methods are classified either according to the type of mobile and stationary phases utilized or according to the mechanism of retention. In referring to a specific type of chromatography, both the mobile and the stationary phases are stated, with the moving phase always referred to first. For example, in gas–liquid chromatography, the mobile phase is a gas and the stationary one a liquid. Even though the liquid phase is bonded to a stationary support, it is still referred to as a liquid stationary phase. In the literature, however, if only one phase is mentioned, the mobile phase is used, such as the shortened terms, gas chromatography or liquid chromatography. In the first case the moving phase is a gas and in the second a liquid. The branches of chromatography are outlined in Fig. 1–1.

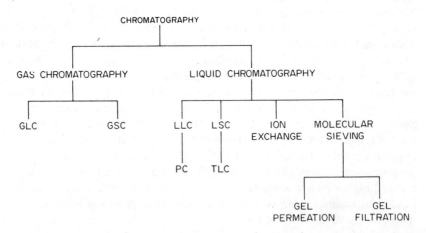

Fig. 1–1. Branches of chromatography.

Therefore, liquid chromatography is that chromatographic process in which the moving phase is a liquid which percolates over a stationary bed. If an open tube or column is used and the technique performed in vertical columns by gravity flow, the method is referred to as classical or conventional column chromatography. Liquid chromatography can also be performed on an open bed, as in paper or thin layer chromatography. High pressure liquid chromatography is a liquid chromatographic technique in which the solution is pumped through the column at pressures up to and sometimes exceeding 4000 psi inlet pressure. The stationary phase may be solid or liquid, but must have a large surface area. If the stationary phase is solid, the technique is called adsorption chromatography. If the stationary phase is a liquid, the sample is partitioned between the stationary and mobile liquid. This process is referred to as liquid–liquid or partition chromatography. Normally, the stationary liquid is polar and the mobile liquid nonpolar. This combination is most commonly used. The liquid chromatographic method using a nonpolar stationary phase and a polar mobile phase is termed reversed phase partition chromatography and it has been found useful in many applications. If the stationary phase is a highly polar, ion exchange resin, the procedure is referred to as ion exchange chromatography. The chromatographic separation takes place by an exchange of ions between the mobile phase and the resin. Compounds having different affinities for the resin will be retained on the column for different periods of time.

Another class of liquid chromatography known as molecular sieving, exclusion chromatography, or gel permeation chromatography (GPC) involves the mechanical sorting of molecules based on the size of the molecules in solution. Separation is achieved with a porous packing gel which is compatible with the mobile solvent.

A. Liquid–Liquid Chromatography (LLC)

In solution or partition chromatography, the sample is partitioned between the mobile and stationary liquids. The main requirements for this type of chromatography are that the solutes must be soluble in the mobile phase and that the mobile liquid not be a solvent for the stationary phase. An example of high pressure liquid chromatography in which the stationary phase is a liquid is illustrated in Fig. 1–2A in which the solid support is coated with 2.0% β,β'-oxydiproprionitrile. Figure 1–2B illustrates reversed phase partition chromatography. It is interesting to note that one of the most widely used subgroups in LLC is paper chromatography in which the more polar stationary phase is held on the cellulose fibers. The mechanism, however, is not just partition, but more complex.

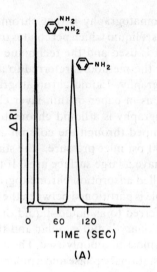

Fig. 1–2. (A) Partition chromatography (liquid–liquid). *Instrument*: Waters ALC-201 liquid chromatograph. *Column*: 50 cm × 2.3 mm i.d.; packed with 37–50 μm Corasil II. *Liquid phase*: 2% β,β'-ODPN. *Eluent*: n-heptane. *Flow rate*: 3.0 ml/min. *Sample*: 5 μl aniline and o-aminoaniline. (Contributed by Waters Associates, Inc.)

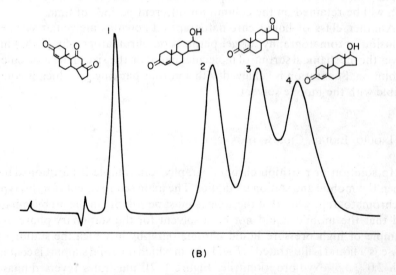

Fig. 1–2. (B) Reversed phase liquid–liquid chromatography. *Instrument*: Varian 4100 liquid chromatograph. *Column*: 75 cm × 2.4 mm i.d.; packed with Permaphase ODS. *Eluent*: 10% MeOH/90% H_2O. *Flow rate*: 30 ml/hr. *Pressure*: 130 psi. *Chart*: 20 in/hr. *Separation*: 2μl. *Samples*: (1) 0.2 mg adrenosterone; (2) 0.3 mg 19-nortestosterone; (3) 0.4 mg 4-androstene-3,17-dione; (4) 0.5 mg testosterone. (Contributed by Varian Associates 1971).

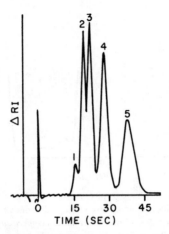

Fig. 1–3. Adsorption chromatography (liquid–solid). *Instrument*: Waters ALC-201 liquid chromatograph. *Column*: 50 cm × 2.3 mm i.d.; packed with 37–50 Corasil I. *Eluent*: *n*-hexane. *Flow rate*: 3.0 ml/min. *Sample*: (1) Aldrin impurity: (2) Aldrin (6 $\mu g/\mu l$); (3) *p,p'*-DDT (6 $\mu g/\mu l$); (4) DDT (8 $\mu g/\mu l$); (5) Lindane (10 $\mu g/\mu l$). (Contributed by Waters Associates, Inc.)

B. Liquid–Solid Chromatography (LSC)

In adsorption chromatography, adsorbents such as silica gel, alumina, or porous glass are used as the stationary phase. The components of the sample, which are adsorbed on the column, are displaced at different rates, depending upon their affinity for the solid. An illustration of a chromatogram in which the stationary phase is the solid, Corasil I, is shown in Fig. 1–3. In thin layer chromatography, which is a subgroup of LSC, the solid phase is spread in a thin film on a glass or plastic sheet instead of being in a column.

C. Ion Exchange Chromatography

Ion exchange chromatography, which is in reality a subgroup of liquid–solid chromatography, has been used extensively in the separation of amino acids and more recently in the separation of nucleic acid components. The stationary phase, which is a solid polyelectrolyte or ion exchange material, has fixed charges, each having an associated counterion of the opposite charge. The ionic solute distributes itself between the mobile liquid and the stationary phases by exchanging ions with a counterion of the solid electrolyte as it passes through the column. Since various ionic solutes have different

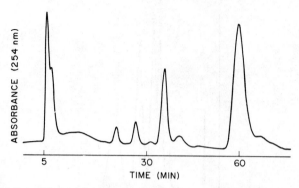

Fig. 1–4. Ion exchange chromatography. *Instrument*: Varian LCS 1000. *Column*: 1 mm i.d. × 3 m; packed with pellicular anion exchange resin. *Eluents*: 0.015 *M* KH$_2$PO$_4$; 0.25 *M* KH$_2$PO$_4$ in 2.2 *M* KCl. *Flow rates*: 12 ml/hr, 6 ml/hr. *Temperature*: 75°C. *Sample*: nucleotide extract of dog blood. [Contributed by P. R. Brown *et al.* (1972).]

affinities for the exchange site, the solutes pass through the column at different rates. However, adsorption on the matrix and partitioning between mobile and stationary phases can also play a role in the retention of the solutes on the column. The major advantages of ion exchange chromatography are: (1) good separation of polar and ionic compounds can be obtained; (2) ion exchange resins are available in high performance forms; and (3) the column packings are very stable and do not degrade easily or quickly. A limiting factor in using ion exchange chromatography is that work with ion exchangers is normally done with aqueous solutions. Usually, compounds must have an ionizable, functional group; however, sugars can also be separated with this technique. Desirable properties of ion exchange materials for chromatography are that they should have low resistance to fluid flow and a high exchange capacity. An example of a chromatogram in which ion exchange chromatography was used is that of an extract of nucleotides of dog blood as shown in Fig. 1–4 and a schematic diagram of the process is shown in Fig. 1–5.

D. Exclusion Chromatography

In exclusion, gel filtration, or gel permeation chromatography the separation of the compounds is according to the molecular size of the components. The stationary phase is a highly porous, nonionic gel in which the size of the pores is similar to the dimensions of the component molecules. The smaller molecules are able to penetrate the gel, whereas the larger ones are not. Therefore, the separation is according to molecular weight or size with the

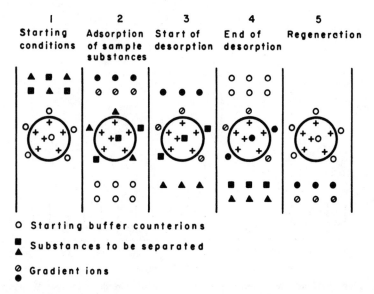

Fig. 1-5. Schematic diagram of ion exchange chromatography. (Contributed by Chromatronix, Inc.)

largest molecules having the shortest retention time. Since the smaller molecules enter the polymer network, their progress through the column is impeded and they have longer retention times than larger molecules. In contrast to liquid–liquid, liquid–solid, and ion exchange chromatography, generally one unique compound is not separated from another. This process is known as fractionation and is usually undertaken to determine molecular weight distribution as an aid in characterizing polymeric compounds. It is also used to produce fractions of narrow, definable molecular weight distributions of mixtures of large molecules for further research. A systematic study of the major factors which influence fast analysis by gel permeation has been made by Little *et al.* (1969, 1970b). A typical chromatogram using this process is shown in Fig. 1–6 and a schematic diagram is shown in Fig. 1–7.

Although exclusion chromatography has been used primarily for high molecular weight compounds, Bombaugh *et al.* (1969) demonstrated the applicability of this technique to the analytical separation of low molecular weight compounds as discrete species. High resolution of triglycerides was accomplished in two ways. The first employed a long column (160 ft) which yielded 180,000 theoretical plates. The second utilized a recycling technique with relatively short columns (16–20 ft) to achieve high resolution. The recycle technique has also been used by Bombaugh and Levangie (1970) to resolve discrete species of higher molecular weight compounds.

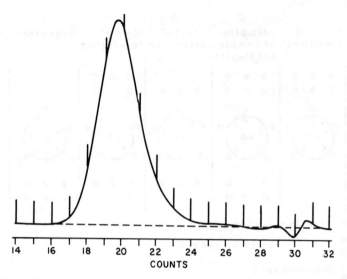

Fig. 1–6. Gel permeation chromatogram of polyvinyl chloride. *Instrument*: Waters ALC/ GPC-301. *Column*: 12 ft × 0.0303 in. *Support*: Styragel, 3 ft-$(10)^6$ Å, 3 ft-$(10)^5$ Å, 3 ft-$(10)^4$ Å, 3 ft-$(10)^3$ Å. *Carrier*: THF, 1.0 ml/min. *Detector*: RI. (Contributed by Waters Associates.)

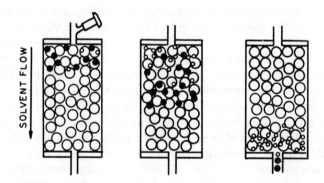

Fig. 1–7. Schematic diagram of gel permeation chromatography. (Contributed by Chromatronix, Inc.).

To get the best results from liquid chromatographic systems, it is important to choose the correct mode for the analysis desired. The first step in an analysis of a mixture of components in a solution is to determine the range of molecular weights and to separate the fractions by molecular size using gel permeation chromatography. With the aqueous solutions, it must then be determined whether the species are ionic or nonionic. If nonionic, then

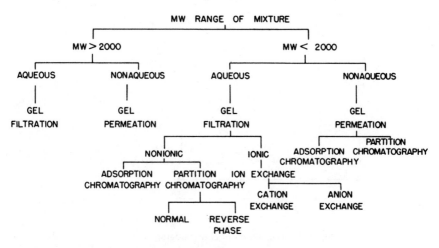

Fig. 1–8. Choosing correct mode of liquid chromatography.

either adsorption or partition chromatography can be used to separate, characterize, and quantitate solutes of each fraction. If, however, the species are ionic, ion exchange separation is the mode of chromatography to be used, and it must be determined whether cation exchange or anion exchange should be utilized. This is graphically portrayed in Fig. 1–8.

Basically, the apparatus of a high pressure liquid chromatographic system consists of a column, a hydraulic system, a detector, and a recorder. If a strip recorder is used, the written record from the chromatographic analysis is a chromatogram. More sophisticated models also include an integrator with printed readouts or an on-line computer in which case the areas are given in numerical values along with retention times.

The term retention time is the time, t_R, from the injection time to that at which the maximum of the solute peak appears on the recorder chart paper. When a component is not adsorbed on the stationary phase, it appears at point A on the chromatogram. This is referred to as the holdup time, t_0, or the void volume V_0. That is, the holdup time or void volume is the volume of the column not occupied by the packing and is equal to the total volume of solvent eluting from the column between the time of injection and the appearance of the unadsorbed species. The adjusted retention time of a solute is defined as the retention time minus the holdup time ($t_R - t_0 = t_A$). The base of the solute peak (B) is determined by drawing tangents to the slope at the inflection point of the peak and determining the distance between the place where the tangents intersect the baseline. The height (H) is calculated from the peak maximum to the baseline. These terms are illustrated in Fig. 1–9.

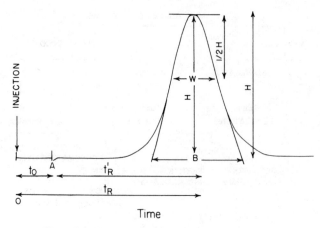

Fig. 1–9.   Illustration of chromatographic terms.

For effective liquid chromatographic separations, a column must have the capacity to retain samples, the ability to separate sample components, and efficiency of operation. The capacity factor of a column is a measure of sample retention by the column and is defined by the expression:

$$k'_R = \frac{t_R - t_0}{t_0} = \frac{V_R - V_0}{V_0}$$

where the values t's are retention times and V's are the retention volumes. If k' values are too low, the components elute too quickly. Therefore, the solvent is too strong. Conversely, if the k' values are too high, the elution times are too long. The long retention times can usually be decreased by increasing solvent strength.

A separation factor, α, which is a measure of a column's ability to separate two components is expressed as a ratio of the capacity factors for the two components 1 and 2.

$$\alpha = \frac{k'_1}{k'_2}$$

If α is 1, the peaks coincide and there is no separation.

In gas and liquid chromatography, the term resolution is used to denote both column and solvent efficiencies. It accounts for both the narrowness of the peaks and the separation of the maxima of two peaks. Column efficiency is concerned with the peak-broadening of an initially compact band as it passes through the column. The broadening is related to column design and operating conditions. It is quantitatively described by the number of theoretical plates, N. To calculate the number of theoretical plates, the equation

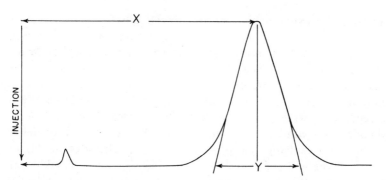

Fig. 1–10. Illustration of terms used to describe number of theoretical plates.

$N = 16 (x/y)^2$ is used, where x equals the distance from injection to peak maximum and y is the length of the baseline of the peak as delineated by two tangents (Fig. 1–10). In order to compare column efficiencies by this method, all operating parameters such as mobile phase, solute, temperature, flow rates, and sample size must be the same. For good resolution, narrow baseline widths (minimum band spreading) are desirable. Therefore, the narrower the peak, the higher the N value and the more efficient the column. Another column efficiency measure is the height equivalent to a theoretical plate, HETP, which is obtained by dividing the length of the column, L, by the number of theoretical plates:

$$\text{HETP} = L/N$$

II. History

The need for instrumentation to separate and determine nucleotide concentration has been evident to those working in the biomedical field for many years. With the discovery of DNA, however, it was even more necessary to develop a reliable technique for the separation and determination of subnanomole quantities of nucleotides in hydrolyzates of DNA and RNA. The methods available were either slow and tedious or not accurate enough. Usually, large samples of tissue were required. There were several practical considerations that impeded the development of high pressure liquid chromatographs. First, column packings had to be found that could withstand the high pressures used to achieve the separation. Second, a highly sensitive monitoring system was essential to separate and quantitate compounds on the subnanomole level. Third, a reliable hydraulic system was needed to achieve reproducible results.

In 1941 Martin and Synge first described liquid–liquid chromatography. With the development of ion exchange chromatography Cohn (1949, 1950) was able to separate not only the purine and pyrimidine bases but the nucleosides and nucleotides as well. Moore and Stein (1957) extended the use of ion exchange chromatography to amino acids and automated the analysis. Hamilton (1962, 1963, 1965) improved the speed of resolution of the amino acid separation, while Crampton *et al.* (1960) introduced the concept of a gradient elution with acidic salt solutions. In 1963, Anderson *et al.* developed automated ion exchange chromatography for nucleotides similar to that used for amino acid analysis. A few years later, Green *et al.* (1966), Huber and Hulsman (1967), R. P. W. Scott *et al.* (1967). Horvath *et al.* (1967). Horvath and Lipsky (1969), Snyder (1967), Uziel *et al.* (1968), and Kirkland (1968) advanced the development of liquid chromatography by working with fine particle size packings, narrow bore columns, and high column inlet pressures. Kirkland and DeStefano (1969), Kirkland and Felton (1969), Horvath *et al.* (1967), Horvath and Lipsky (1969), Burtis and Gere (1970), and Brown (1970) defined a wide range of operating parameters and by optimizing the conditions made possible rapid, quantitative determinations of subnanomole quantities of bases, nucleosides, and nucleotides. The development of small particles in the new packings, such as controlled surface porosity anion and cation exchange packings by Kirkland and his group at du Pont (1969a), the pellicular ion exchanger by Horvath and co-workers at Yale (1967), and Corasil and Porasil by the group at Waters Associates (1970) as well as the use of narrow bore columns greatly improved efficiency and reduced analysis time.

Although many detectors have been tried, the most sensitive detector currently used in high pressure liquid chromatography is a micro-UV detector. In 1961, Anderson *et al.* (1963) adapted a continuous column effluent monitoring technique for use in his system. The UV detector with adsorbance at 254 nm was further modified and used by C. D. Scott (1968), Horvath and Lipsky (1969c), Horvath *et al.* (1967), Kirkland (1969a), and Felton (1969). Detectors which monitor the refractive index of the effluent are also widely used. A detector measuring heat of adsorption is now being marketed. The feasibility of detectors using other physical properties is under intensive investigation. At present 95% of all detectors used utilize either refractive index or UV absorption.

It has only been within the last few years that high pressure liquid chromatographs have been commercially available. Now, not only are these instruments being manufactured by at least six companies, but components are readily available from these companies and several others so that noncommercial instruments can be built.

III. Comparison of Chromatographic Methods

Although classical column chromatography is the oldest chromatographic method, gas chromatography and thin layer chromatography have dominated the field for the past decade. The advantages and disadvantages of different types of chromatography were well summarized by Huber (1969b). In the last decade gas chromatography has been the method of choice for many reasons. The same column can be used for many and different separations, and a high plate number per unit of time may be obtained. The precision of the quantitative analysis and retention data for identification is good. It can be used on-line with other instrumental methods, for example, mass spectrometry. It can also be automated. However, the requirement that the sample must be volatile makes its application limited, especially in the biomedical and biochemical areas, where most of the compounds encountered are nonvolatile, polar, or thermally labile. Thin layer chromatography, on the other hand, has the advantages of requiring no expensive instrumentation and it has the capability of separating a number of samples simultaneously. Both the stationary phase and the moving phase can influence the separation. If two-dimensional thin layer chromatography is used, the separation can be improved by using two different eluents. However, neither the quantitative composition data nor the migration data have a high degree of precision and it is difficult to automate. Moreover, the number of theoretical plates is limited. Therefore, there was a need for a high speed, efficient liquid chromatographic technique to obtain in a shorter time better separation of highly reactive compounds which are not suitable for analysis by gas chromatography. Although it was previously assumed that liquid chromatography is an inherently slower process than gas chromatography, it has now been shown that liquid chromatography can be very fast, especially for simple separations. Waters *et al.* (1969) found that the length of time for a chromatographic separation was a direct function of the resolution required and separation factors available. After examining both published references and new data, they found that the classical Van Deemter equations must be expanded for application to high speed liquid chromatography. It was demonstrated by Gordon (1963), Giddings (1965), and Knox (1966) that additional factors are necessary to account for coupling and turbulence. Huber (1969) found that at least four terms are necessary to describe the plate height in the Van Deemter equation. These are longitudinal molecular diffusion (H_{MD}), dispersion caused by variations of fluid velocity in the intergranular pores (H_{MC}) and H_{EM} and H_{ES}, which are terms which depend on the speed of mass transfer in the mobile and stationary phases, respectively.

$$H = H_{MD} + H_{MC} + H_{EM} + H_{ES}$$

IV. Advantages of High Pressure Liquid Chromatography

In order to obtain excellent high speed chromatographic separations, it was found by the pioneers in the field that small bore columns packed with small size particles gave the best results. This necessitated the use of high inlet pressures to maintain high flow rates through this bed of fine particles. Hence, the terminology high speed, high performance, high efficiency, and high pressure liquid chromatography. High pressure liquid chromatography has many advantages over conventional column chromatography. It is fast and versatile. It is simple to operate and requires relatively little skill. It has high sensitivity (in the picomole to nanomole region), gives good resolution, and affords both qualitative and quantitative analyses. Furthermore, the sample is not destroyed and fractions can be collected. Relative to previous techniques, it is fast; complete analysis of the nucleotides in a cell extract can be obtained in 1 hour. Within a few minutes, the content of a drug or its metabolite in a physiological fluid can be determined. With its sensitivity, nanogram (10^{-9} gm) quantities of material can be measured. Also the sample size requirements are small and usually a complete analysis can be carried out on a few microliters of sample. The chromatograms are reproducible with dependable retention times and accurate results. A wide range of compounds can be separated by high pressure liquid chromatography; for example, free nucleotides, nucleosides, or bases or these compounds from RNA hydrolyzates, drugs and their metabolites, steroids, pesticides, vitamins, alkaloids, amino acids, flavors, foods, antioxidants, plasticizers, lipids, and proteins. Impurities in pharmaceuticals can also be detected by this method. In fact, almost any compound that is soluble can be separated by liquid chromatography. High pressure liquid chromatography can be used for studies in chemistry, biochemistry, biochemical pharmacology, biology, pathology, physiology, nutrition, and, clinically, in metabolic studies, in nutritional studies, and in diagnostic tests. It is possible to follow drug metabolism in any of the physiological fluids.

High pressure liquid chromatography has many advantages over gas chromatography. In gas chromatography, the solute must be volatile and thermally stable, whereas in liquid chromatography, high molecular weight compounds and those that are labile, polar, or nonvolatile can be separated. The main requirement is that the solute be soluble in the mobile solvent. Although in gas chromatography the efficiency is good, ~ 2000 plates/m, in liquid chromatography, the efficiency is even better, ~ 5000 plates/m. It is possible to vary the stationary phase, the flow rate, and the temperature in gas chromatography; in liquid chromatography, not only can these variables be changed, but also the mobile phase can be varied. Either isocratic elution or gradient elution can be used and the gradient can be linear, concave, or convex. Moreover, different salt concentrations, salt anions, and pH of the eluents can be used. With liquid–liquid chromatography, organic solvents may be used as the eluents instead of salt solutions. High pressure chromatography is more quantitative than thin layer chromatography and the results can be as accurate and precise as those obtained in gas chromatography.

Another advantage of high pressure liquid chromatography is that the columns are reusable. Since the stationary phase is usually stable, and since the mobile phases are chosen so that they will not alter the characteristics of the columns, with proper care the columns should have a long life span.

Even though high performance liquid chromatography systems may use high pressures, they are not dangerous to use. Unlike gases, liquids are not as compressible and very little potential energy is stored in a liquid under pressure at ambient temperature; thus, there is no danger of explosions.

CHAPTER 2

||

INSTRUMENTATION

The liquid chromatography systems on the market today are usually referred to as high pressure liquid chromatographs. Although up to 5000 psi may be used, efficient operation is obtained many times at much lower pressures, such as 100–200 psi. The results which can be obtained by high pressure liquid chromatography, as compared to conventional chromatography, are illustrated in Fig. 2–1. In the high pressure system, complete resolution was obtained in 20 minutes using a packing with a particle size of 10 μm, pressure of 400 psi, and a flow rate of 2 ml/min. Incomplete separation was obtained in a conventional gravity flow system and the separation took 3 hours and 10 minutes. The particle size was 100 μm, the flow rate 0.2 ml/min, and the pressure was 0.5 psi.

The essential components of a high speed liquid chromatographic system are a hydraulic system, a column, and a detector. Also, temperature controls and good sample introduction devices are necessary. The development of an efficient liquid chromatographic system, which has the speed, resolving power, and sensitivity of gas chromatography, was hindered by the lack of a good, leak-free hydraulic system capable of operating at high pressure. A new approach to column design was necessary to produce high column efficiency, and to improve resolution and reproducibility. Most important of all was a sensitive, low dead-volume detector. After the initial work of Anderson (1962), little attention was paid to liquid chromatography because components were not readily available. R. P. W. Scott et al. (1967) described some factors that affect the design of a liquid chromatograph. Horvath et al.

16

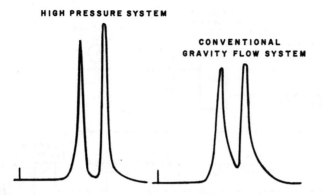

Fig. 2–1. Comparison of chromatograms of high pressure liquid chromatograph and conventional gravity flow system.

	High pressure	Conventional
Resolution	Complete	Incomplete
Speed	20 min	3 hr and 10 min
Particle size (approx.)	10 μm	100 μm
Flow rate	2 ml/min	0.2 ml/min
Column length	9 in.	9 in.
Column diameter	$\frac{1}{4}$ in.	$\frac{1}{4}$ in.
Sample size	0.5 ml	0.5 ml
Pressure	400 psi	0.5 psi (14 in. water)

Conditions common to both separations: Sample was 0.5 ml of about 0.5 mg/ml each of sodium indophenol (second peak) and methyl orange dissolved in 15% methanol in water solvent. Mobile phase solvent was 15% methanol in water. In the "conventional" system the stationary phase was 100–200 mesh silica gel. In the high pressure system the stationary phase was the same silica gel which had been ground and passed through a 325 mesh screen. Note that these curves are both plotted on a volume basis, not on a time basis. This allows direct comparison of the resolution. (Contributed by Chromatronix, Inc.)

(1967) presented their design for a liquid chromatographic system, which featured high inlet pressures, a sensitive UV detector, and novel pellicular material for column packing. An instrument based on this concept was originally built and sold by Picker Nuclear and is now marketed by Varian Aerograph. A simplified flow diagram is shown in Fig. 2–2. Huber and Hulsman (1967), in their study of liquid chromatography in columns, reported on a similar chromatographic apparatus, which they described as consisting of two major systems: a separation and a measurement system. This chromatograph had been in use in their laboratory for more than 2 years. A flow diagram of their system is shown in Fig. 2–3. Snyder (1967) described a system which he used for high efficiency separation by liquid–solid chromatography. In

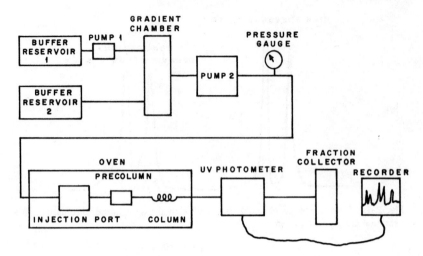

Fig. 2–2. Functional diagram of Varian LCS 100 liquid chromatograph.

1968 Halász *et al.* and Jentoff and Gouw (1968) also reported on their designs of modern liquid chromatographs. Felton in 1969 outlined the components of a liquid chromatographic system that was developed and is now marketed by du Pont. A flow diagram of the du Pont 820 liquid chromatograph is illustrated in Fig. 2–4. Bombaugh *et al.* (1970) examined the critical components and presented a design of a simplified, high pressure liquid chromatograph which is marketed by Waters Associates (Fig. 2–5). A diagram of the Perkin/Elmer liquid chromatograph is shown in Fig. 2–6. An instrument proposed by Laboratory Data Control to be put together by the researcher is shown in Fig. 2–7, and one that can be built from standard hardware and components available from Waters Associates is shown in Fig. 2–8.

All instruments that have been discussed so far are equipped with a single column. Varian now offers an instrument with dual columns to compensate for any shift in baseline which might possibly be caused by the gradient eluent. The system was constructed so that columns with similar flow characteristics were connected in parallel, and volumes from the splitting tee (before the injector) to the fluid cells were identical. This setup is especially useful in complete urinalysis, which may take as long as 15 hours. A flow diagram of the Varian dual column LCS 1000 is shown in Fig. 2–9. Columns also may be used in series. For example, it is possible to use a strong and a weak ion exchanger in series to achieve difficult separations. C. D. Scott *et al.* (1972) described a liquid chromatographic system in which they used anion and cation exchange columns both in series and in parallel to obtain an increase in resolution, a decrease in separation time, and a high degree of automation in the analysis of complex mixtures of

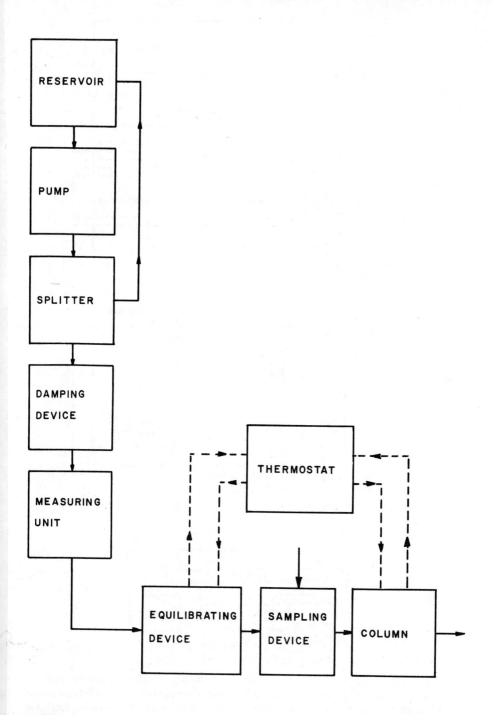

Fig. 2–3. Huber's block diagram of separation part of an earlier modern liquid chromatograph. [Contributed by Huber (1969a).]

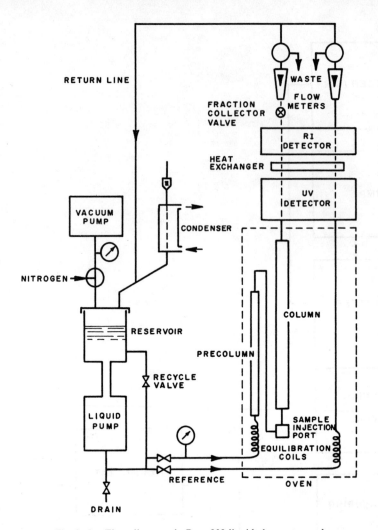

Fig. 2–4. Flow diagram du Pont 820 liquid chromatograph.

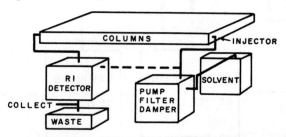

Fig. 2–5. Flow diagram of Waters ALC 201. Solvent is pumped from the reservoir, through the on-column injector to the column system for separation. The separated sample components then pass through the detector where the solute bands are sensed as a difference in refractive index. The resulting difference signal is transmitted to a recorder. Eluent then may be either collected or deposited in the waste reservoir. (—) Sample; (--) reference.

20

molecular constituents in body fluids. With proper solvent programming it is not difficult to imagine a setup in which columns packed with nonpolar materials are linked in series to those with anion and cation exchange resins to separate a whole spectrum of different compounds.

A new technique which is suited to high efficiency repetitive analysis, or where increased resolution is required at certain points, is column programming. It is also referred to by Snyder (1969b) as coupled-column opera-

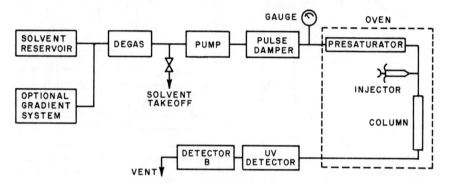

Fig. 2–6. Flow diagram of a Perkin/Elmer 1240 liquid chromatograph.

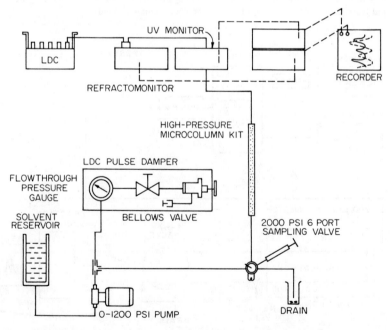

Fig. 2–7. Flow diagram of noncommercial liquid chromatograph with components and design supplied by Laboratory Data Control. (—) Fluid; (--) electrical.

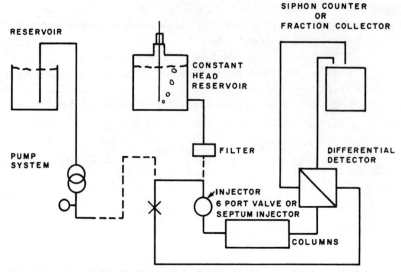

Fig. 2–8. Typical liquid chromatography system composed of standard laboratory hardware and components from Waters Associates. In the upper middle is a gravity feed system, consisting of a constant head reservoir and a filter. In the upper left-hand corner is a pumped feed system consisting of reservoir and pumping system. One of these alternative solvent feed systems is used and it is followed by an injector, column, detector, siphon counter, and fraction collector, as is appropriate for your particular needs.

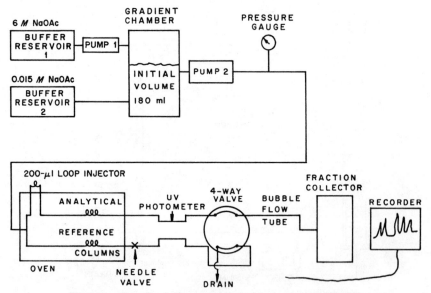

Fig. 2–9. Functional diagram of Varian dual column LCS 1000.

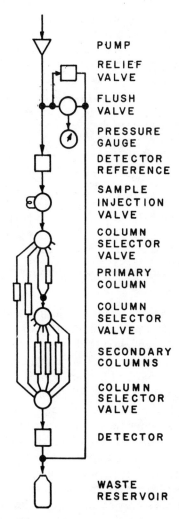

PUMP

RELIEF
VALVE

FLUSH
VALVE

PRESSURE
GAUGE

DETECTOR
REFERENCE

SAMPLE
INJECTION
VALVE

COLUMN
SELECTOR
VALVE

PRIMARY
COLUMN

COLUMN
SELECTOR
VALVE

SECONDARY
COLUMNS

COLUMN
SELECTOR
VALVE

DETECTOR

WASTE
RESERVOIR

Fig. 2–10. Flow diagram of Chromatronix coupled column system.

tion or "multiple column chromatography." After a short, primary column provides initial separation, a valve directs the effluent to one or more secondary columns. The Chromatronix coupled column system schematic flow diagram is shown in Fig. 2–10. Because of the use of three column selection valve modules, two standard (noncoupled) columns can be used. Difficult separations, however, may preclude the use of multiple column chromatography or dual or multiple detectors for analysis of a complex mixture. Another approach to tackling the problem of separating, identify-

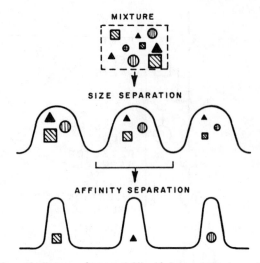

Fig. 2–11. Schematic diagram of sequential liquid chromatography analysis. (Contributed by Waters Associates, Inc.)

ing, and measuring the components in a solution is sequential analysis by liquid chromatography. In this technique, two or more modes of liquid chromatography are used sequentially. For example, the solution is first analyzed by gel permeation for size separation of the particles. The collected fractions are then further analyzed by partition or adsorption chromatography and then, if necessary, by ion exchange chromatography (Fig. 2–11).

Still another method to increase effectively column length and obtain better resolution is the recycling technique in which the effluent is recycled into the column. Until now this technique has been used primarily in gel permeation chromatography. This system is less expensive than a series of columns because it requires less packing material and is possible to use because the plate numbers of equivalent columns are additive. To overcome the high cost and high pressures required with the long columns needed for high resolution in gel permeation chromatography, a recycle technique has been developed by Bombaugh and Levangie (1970). They found that a recycling operation through the reciprocating pump, gel permeation column, and detector provides high resolution on both analytical and preparative scale.

Two other companies that market both complete high pressure liquid chromatographic systems and components are Chromatronix and Chromatec, Inc. All combined instruments feature a good solvent supply system, long compact columns with narrow bores packed with microspherical hard core particles, and sensitive detectors with small dead-volumes. The instruments vary somewhat in features offered, design, and possible applications, although the basic components are similar. Each researcher must decide which instrument is best for his particular needs, what features are most needed for his work, and how the instrument is to be used. Extra features add to the expense and are not necessary if only one or a series of analyses are to be carried out routinely. However, for other workers who are going to use the chromatograph for a variety of projects, versatility may be most important. It is possible to buy separate modular units or components to build your own instrument at lower cost. This has been done satisfactorily by Anderson (1962), R. P. W. Scott *et al.* (1967), Huber and Hulsman (1967), Felton (1969), Anders and Latorrè (1970), Bombaugh *et al.* (1970), Siggia and Dishman (1970), and others. Companies such as Laboratory Control Data and Analabs sell components and Perkin/Elmer, formerly Nester/Faust, Waters Associates, Chromatronix, and Chromatec market liquid chromatography stock parts to assist those who wish to build their own systems. The major advantage in building your own system is lower cost. It is also easier to optimize the components for a particular separation problem. The main disadvantages are the time needed (which can cut into valuable research time), the fact that no service, and sometimes the finer controls are not available. Here again, each researcher must weigh the disadvantages and advantages and come to conclusions based on his time, resources, and talents.

I. Hardware

Although hardware generally includes pumps and detectors, these important parts of a high pressure liquid chromatographic system are discussed under separate sections later in this chapter. Since a fast analysis, high resolution, and high sensitivity are required of a liquid chromatographic system, the hardware must have maximum resistance to corrosion. For good quantitation there must be a high degree of reproducibility. The hardware should be capable of handling high inlet pressures and high flow velocities. It is necessary for good resolution that all components, plumbing connections, injector, and detector have a low dead-volume. Since solvent flow rates must be carefully controlled, Chromatec now offers an automatic flow measuring device connected to an event marker on the chart which records the rate on the chart. This is important because reproducible quantitation depends on accurately controlled flow rates whose precision should be within 2.0%. It was found by Brown (1971a) that flow rate variation not only affects elution time and peak shape but also accurate quantitation. Controls must also be available for precisely regulating the flow rate of the high concentration buffer into the low since this flow rate also influences the retention data. When a gradient is used, a reservoir that is easily and quickly filled is necessary for the low concentration eluent. A multifunctional gradient device has been described by Byrne and Schmit (1971) which allows the generation of a variety of linear and exponential gradients as well as stepwise elution. According to the authors, this time-proportioning electromechanical system controls gradient shapes, total gradient times, and the initial and final gradient concentrations without equipment changes or calculations.

It is helpful to have a method for degassing the solvent since air bubbles will block the column, and for high sensitivity detection it is necessary for the mobile phase to be free of dissolved gas. Other options that may be included in the reservoir are a heater, stirrer, and inlets for vacuum or nitrogen purge since degassing may be accomplished by pulling a vacuum on a filled reservoir for a few minutes while stirring the eluent vigorously. Heating or a nitrogen purge may be necessary to remove dissolved oxygen from some solvents.

Because reproducible and accurate results are necessary, column temperature must also be controlled. Some systems include circulating air bath column oven. Others use a thermostated circulating water bath. The major advantage of the air bath is that air can be circulated very rapidly so that the air oven can do a better job of temperature control.

Since the sample is not destroyed with the majority of detectors used in high pressure liquid chromatography, a fraction-collecting valve is valuable to collect fractions of the effluent containing chromatographic peaks. This can be accomplished by manual collection or by the use of commercially available automatic continuous fraction collectors. A simple three-way valve attached to the detector outlet can be used to divert the mobile phase to collection vessels for preparative work or additional analytical investigations.

There are several techniques for the injection of the sample that can be used with high performance liquid chromatography. One method is the direct or live injection of the sample directly into the column or pre-column by a precisely calibrated syringe. During the injection the pressure is stopped so that it is very easy to inject the sample completely. Another method is the use of an on-column septum injector which permits injection into the flowing stream at pressures less than 1000 psi. If pressures are greater, the stop-flow input technique must be used. In these manual methods, low dead-volume injections are possible which provide high efficiency. The loop or sample valve injector provides a convenient method of injecting either small or large sample volumes and can be used for preparative scale applications. It is also useful for reproducible sample injection for routine, repetitive operations. If a gradient elution is not used, the injection procedure for a high pressure liquid chromatographic system can be automated with good precision using the sample valve devices. Although automatic sampling systems for gel permeation chromatography have been utilized for some time, automatic sampling devices for other liquid chromatographs are just coming of age. Sample valves which are operated mechanically are now available from Valco Co. (Houston, Texas), and the Hamilton Co. (Whittier, California) has developed a pneumatically operated valve for these systems. Chromatronix, Inc. has available many sample injection valves, including an automatic sample injection valve, an automatic stream sample injection valve, a recycling valve, and an automatic recycling valve as well as sample loops for samples ranging in volume from 1.00 to 10.0 ml. They also have a 6-position rotary valve and a pneumatic activated valve.

II. Pumps

There are three types of pumps that have been used in liquid chromatography: a diaphragm pump, a positive displacement-type pump, and a pressurized tank. The diaphragm pump produces a pulsing flow. This is a major disadvantage when using a flow sensitive detector such as the microadsorption detector. However, with a UV detector, this is not a serious drawback. Another objection to the use of a pulsating pump is that optimum conditions are difficult to maintain because the flow is not steady. The apparent average flow may vary enough to cause differences in separation. It is possible to use damping systems to reduce pump pulsation effects, with a resultant steady flow. This may introduce considerable dead volume into the liquid carrier system. Bombaugh *et al.* (1970), however, successfully used a specially designed damper system to maintain a pulse-free operation.

Both the positive displacement pump and the pressurized tank are pulseless and have limited reservoirs. The positive displacement pumps of Nester/Faust and Varian are constant flow and the duPont hydraulic pump is a constant pressure system. In one type of positive displacement pump a piston moves through a certain length of pipe or tube. A modification of this pump is that used by Felton (1969) in which a large pneumatic piston drives a smaller one. This pump delivers 3200 psi and about 60 ml of carrier liquid per stroke. In the pressurized tank, the solvent is forced from the reservoir by applying gas pressure on the surface causing the liquid to flow up through a dip-tube. One of the main problems with this type of pump is that gas dissolves in the carrier liquid. The bubbles in the mobile phase create noise in the baseline or may disrupt the packing in the column.

III. Detectors

A chromatographic detector is a device which indicates the presence of and measures the amount of separated components in the column effluent. An ideal liquid chromatography detector has versatility, high sensitivity, continuous monitoring of the column effluents, low noise level, wide linearity of response, stable baseline, insensitivity to flow rate and temperature changes, and response to all types of compounds. It should also be rugged and not too expensive. The detector should be capable of measuring accurately a small peak volume without appreciably increasing its volume since increases in peak volume decrease the detector's effective sensitivity. Although there are many sensitive, small dead-volume, universal detectors used in gas chromatography, one of the more difficult problems in liquid chromatography is to find an appropriate detector. Since the scope of the separations that can be performed in liquid chromatography is vast, it has been found that there is no true universal detector for all systems. Therefore, it is necessary to select a detector to be used on the basis of the problem at hand. The methods that will be described are effluent stream monitors. For these techniques, conditions such as flow rate, stream temperature, and detector environment are all-important and must be carefully controlled. Conlon (1969) discussed the detectors commercially available and others that showed promise as future monitors. Some detectors used in high performance liquid chromatography are listed in Table 2–1.

The terms noise, sensitivity, and linearity are used in describing detector performance. Noise, which is defined as the variation of the output signal which cannot be attributed to the solute passing through the cell, can result from many causes. Among them are instrument electronics, temperature

TABLE 2–1

DETECTORS USED IN LIQUID CHROMATOGRAPHY

Type of device	Units	Full scale sensitivity at $\pm 1\%$ noise	Sensitivity to favorable sample
Refractive index	RIU	10^{-5}	5×10^{-7} gm/ml
UV absorption	AU	0.005	5×10^{-10} gm/ml
Heat of adsorption	°C	5×10^{-3}	10^{-9} gm/sec
Fluorescence	—	—	10^{-9} gm/ml
Conductivity	μmho	0.05	10^{-8} gm/ml
Polarography	μA	2×10^{-8}	10^{-9} gm/ml
Flame ionization	A	10^{-11}	10^{-8} gm/sec

fluctuation, line voltage change, and changes in flow rates. There are three forms of detector noise: short-term, long-term, and drift. Short-term noise widens the trace and appears as "fuzz" on the baseline, whereas with long-term noise the variation in the baseline appears as "valleys and peaks." The steady movement of the baseline either up or down the scale is referred to as drift.

Sensitivity is divided into two categories: absolute detector sensitivity and relative sample sensitivity. The definition of absolute detector sensitivity is the total changes in a physical parameter for a full-scale deflection of the recorder at maximum detector sensitivity with a defined amount of noise. The relative sample sensitivity of a detector is defined as the minimum concentration of solute which can be detected. In comparing detectors, sensitivity is quoted as the value equal to the noise or to twice the noise. For detectors in high pressure liquid chromatography the signal output should be linearly proportional to the solute concentration. Although most of the detectors approach this desired goal, none are completely linear over their entire range of detectability. Linearity can be measured as the maximum percentage deviation from linearity over a range expressed as a percentage of full-scale deflection.

A. REFRACTIVE INDEX MONITORS

The detector that comes closest to being a universal one is the refractive index detector. The refractive index of the solvent changes whenever a solute is present. This detector is stable and simple to operate and the sample is not destroyed. Although its level of sensitivity is good, it is much less sensitive than a UV detector (1.0 ppm compared to 0.06 ppm). A refractive index detector can be based on one of two principles. In one the angle of deviation of a ray of monochromatic light as it passes through the flowing eluent stream changes as a function of sample concentration. The change in bending angle is monitored by an electromechanical arrangement which moves either the source or a photodetector to maintain optical alignment. This movement is converted into an electrical signal by means of a retransmitting slide wire or balance potentiometer. Another refractive index detector is based on the reflection principle, first elucidated by Fresnel, in which the intensity of the reflected component of a beam of light impinging on the surface of the effluent stream changes inversely with the refractive index. The angle of deviation refractometers are currently available with the same cell volume and at lower cost than the reflection type, although the latter has lower permissible flow. One of the disadvantages of both types of refractive index detectors is that they are difficult to use with a gradient

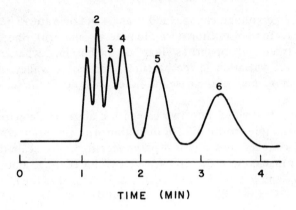

Fig. 2–12. Chromatogram illustrating use of refractive index detector in separating cyclic hydrocarbons. *Instrument*: noncommercial system with Waters R-400 differential refractometer. *Column*: 100 cm × 2.3 mm; packed with Woelm alumina with 6% water added. *Particle size*: 38–53 μm. *Flow rate*: 2.2 ml/min. *Eluent*: n-hexane. *Sample*: (1) decalin; (2) benzene; (3) naphthalene; (4) azulene; (5) O-quaterphenyl; (6) M-quaterphenyl. [Contributed by Bombaugh, *et al.* (1970).]

elution. Also temperature control of the column effluent and detector are critical. An example of a chromatogram obtained using a refractive index monitor is shown in Fig. 2–12 where cyclic hydrocarbons were separated by Bombaugh *et al.* (1970). The refractive index of solvents which can be used in liquid chromatography are listed in Table 2–2.

B. ULTRAVIOLET AND VISIBLE ABSORPTION

When a molecule is exposed to UV radiation, the radiation is absorbed by particular electronic configurations of the molecule. Compounds which contain aromatic rings, $C=O$, $N=O$, $N=N$ functional groups are the principle absorbers in the UV region. Some chromophores that absorb in the UV or visible are listed in Table 2–3 and the UV absorption of chromophoric groups in steroids in Table 2–4. Although extended exposure to ultraviolet radiation may cause photodecomposition of the sample, there is minimum photodecomposition in high pressure liquid chromatographic systems because of the small cell volume, fast flow rates, short exposure, and low intensity of the UV radiation.

TABLE 2–2

RELATIVE UV ABSORBANCE OF SOLVENTS[a]

Solvent	RI	UV cutoff (nm)		
		200	250	300
n-Pentane	1.36			
Petroleum ether	–			
Cyclohexane	1.43			
Carbon tetrachloride	1.47			
Amyl chloride	1.41			
Xylene	1.50			
Toluene	1.50			
n-Propyl chloride	1.39			
Benzene	1.50			
Ethyl ether	1.35			
Chloroform	1.44			
Methylene chloride	1.42			
Tetrahydrofuran	1.41			
Ethylene dichloride	1.45			
Methylethyl ketone	1.38			
Acetone	1.36			
Dioxane	1.42			
Amyl alcohol	1.41			
Diethylamine	1.39			
Acetonitrile	1.34			
Pyridine	1.51			
n-Propanol	1.38			
Ethanol	1.36			
Methanol	1.33			
Ethylene glycol	1.43			
Acetic acid	1.37			
Water	1.33			

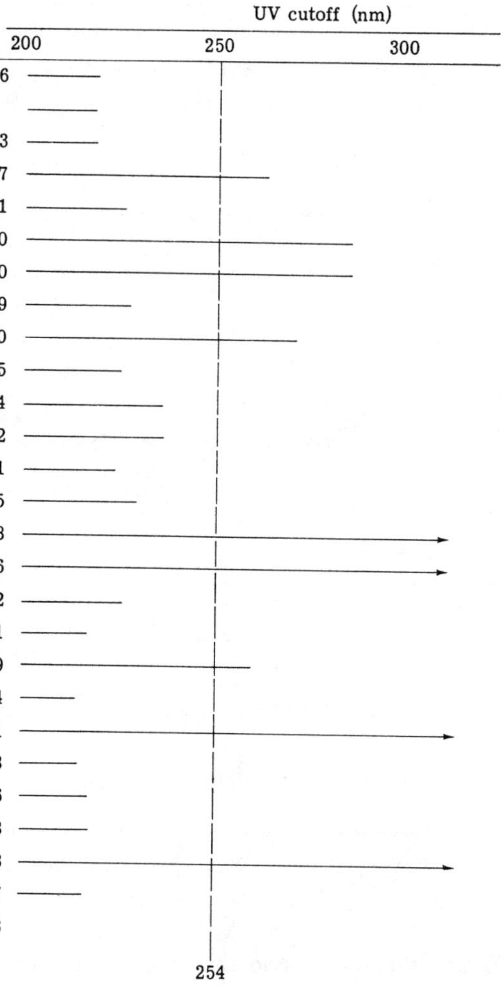

254

[a]Contributed by Perkin-Elmer Corp., formerly Nester/Faust Mfg. Corp.

Even though absorption in the UV at a specific wavelength cannot be universally used as a detector because the sample material must exhibit a suitable extinction in that region of the UV spectrum, its application to stream-monitoring to examine individual portions of the effluent has proved very valuable. Many biochemical compounds absorb to some degree in the

TABLE 2-3

ABSORPTION BAND CENTER FOR DIFFERENT FUNCTIONAL GROUPS[a]

Maximum absorption at less than 230 nm

(1) C—C   (2) C=C   (3) C=C—C=C   (4) C≡C

(5) C=S   (6) S—H   (7) O=C—OH

Maximum absorption between 230 and 270 nm

(1) C=O   (2) 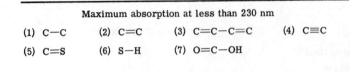   (3) heterocyclics

Maximum absorption between 270 and 320 nm

(1)    (2) $\overset{O}{\overset{\|}{C}}-\overset{O}{\overset{\|}{C}}$ (3) N—N (with N substituents)

(4) N=O   (5) [naphthalene structure]   (6) [benzene ring]—C=C

Maximum absorption greater than 320 nm

(1) N=N   (2) C=C—C(=O)—C=C   (3) [anthracene structure]

---

[a]Contributed by Waters Associates, Inc.

250–260 nm region and this method is especially useful in indicating the presence of purine and pyrimidines and their nucleosides and nucleotides because their molar extinctions are approximately $9-15 \times 10^4$ at this wavelength. Since UV absorption characteristics are generally well known or easy to obtain, this detector is easy to evaluate. Anderson *et al.* (1963) devised one of the first continuous effluent monitoring systems utilizing UV absorbing properties. The fact that a UV detector is sensitive only to those compounds that absorb at a specified wavelength is both an advantage and a disadvantage. The number of compounds that show up in the chromatogram is limited, thereby sometimes clarifying the spectrum so only

TABLE 2–4

Chromophores	Max (nm)	Max	Example	Max (nm)	Max
				Ultraviolet absorption	
Carbonyl	170 – 200 280 – 300	5000 – 10,000 50 – 100	(structure)	282	31
Double bond	180 – 225	500 – 5000	(structure)	204	3,300
a, b Unsaturated ketone	230 – 270	10,000 – 18,000	(structure)	253	11,200
Conjugated diens (different rings)	220 – 260	14,000 – 28,000	(structure)	232 239 248	21,500 23,500 16,000
Conjugated diens (same ring)	250 – 285	5000 – 15,000	(structure)	262 271 282 293	7,700 11,400 11,900 6,900
Conjugated polyenes	280 – 350	5000 – 20,000	(structure)	306	14,500
Dienones (different rings)	280 – 300	10,000 – 30,000	(structure)	284	28,000
Dienones (same ring)	240 – 315	5000 – 15,000	(structure)	244	15,000
Trienones	220 – 380	10,000 – 30,000	(structure)	223 256 298	13,500 11,900 15,300

[a]Contributed by Henry et al. (1971b).

33

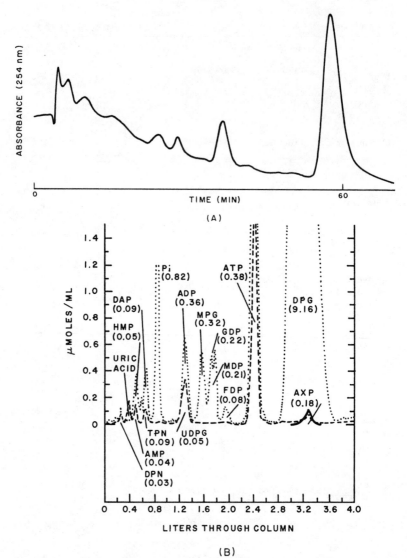

Fig. 2–13. Comparison of chromatograms of human erythrocytes using high pressure liquid chromatography with UV detector and that using column chromatography with detection by phosphorus assay and UV at 260 nm.

(A) *Instrument*: Varian LCS 1000. *Column*: 1 mm × 3 m; packed with pellicular anion exchange resin. *Flow rates*: 12 ml/hr; 6 ml/hr. *Eluents*: 0.015 M KH$_2$PO$_4$; 0.25 M KH$_2$PO$_4$ in 2.2 M KCl. *Temperature*: 75°C. [Contributed by Brown (1970).]

(B) Column chromatography with detection by phosphorus assay and OD at 260 nm. (...) Total P; (--) OD 260; (—) Fe. Numbers in parentheses indicate μ moles P/ml RBC. [Contributed by Bartlett (1970).]

certain compounds show up clearly. The disadvantage, however, is its lack of universality; thus it is not possible to analyze for all the components present in the sample. An example of this is in the analysis of nucleotides in cell extracts in which all other components are either precipitated by the acid or not visible in the UV. This is demonstrated in Fig. 2–13, A and B. In the lower chromatogram of human erythrocytes, the separation was obtained by conventional column chromatography and the measurement of the solutes in the eluent obtained by phosphorus assay as well as by UV determinations (Bartlett, 1970). In comparison, only the nucleotides are visualized in the chromatogram obtained by high pressure liquid chromatography using a UV detector. In the former, compounds such as the 2,3-diphosphates of sugars and 2,3-diphosphoglycerate are present (in addition to the nucleotides), whereas in the latter these are not seen.

Although a UV detector is not universal, it is widely used in liquid chromatography systems because it has a low dead-volume and is relatively insensitive to changes in the eluting solvent when the solvent is a non-UV absorber. The relative UV absorbances of solvents that can be used in liquid chromatography are shown in Table 2–2. It can be used with a pulsating flow and is relatively insensitive to flow rates. It does not need fine temperature controls and can be used with very small sample size, down to the picomole range.

For use in high pressure liquid chromatography, micro-UV detectors have been developed. Radiation from a low pressure mercury tube source which transmits radiation from the 254 nm mercury line has been found to be most suitable because of its intensity and because so many biochemical compounds have some absorbance in that region. The detector most commonly used is a microdetector with a flowthrough absorbance cell with an internal diameter of 1 mm, a small internal volume of approximately 7–12 μl, and a 10-mm pathlength. This detector has low dead-volume and high sensitivity. The range of linearity is good and the sensitivity can be as low as 0.002 absorbancy units. The detector can be adapted for use at other wavelengths. At present, a micro-UV detector which absorbs at 280 nm is available; however, it is not as sensitive as the detector which absorbs at 254 nm. By using different wavelengths, selectivity and sensitivity can be enhanced since a change in wavelength can eliminate the sensitivity to one compound and maintain or increase the sensitivity to another. For example, the purine nucleotides absorb strongly at 254 nm, but their methylmercapto analogs absorb most strongly at 292 nm. Therefore, a shift in wavelength from the 254 nm UV spectrophotometer to 280 nm will increase the sensitivity to these analogs.

If absorption in the visible range is utilized, usually a complexing reagent is used to form a colored derivative since many of the solutes in the effluent are colorless. Although more chemistry is required to develop a suitable reagent for visible absorption, the method is still relatively simple and the ninhydrin reaction has been used routinely in the amino acid analyzer.

C. THE HEAT OF ADSORPTION DETECTOR

The heat of adsorption is utilized in detectors and is universal within certain limits (Munk and Ravel, 1969; Davenport, 1969). It can be used in ion exchange, liquid–solid, liquid–liquid, and gel permeation chromatography. The basis for this detector is the change in temperature with the adsorption phenomenon. Minute changes in temperature, which occur where a solute is adsorbed onto a surface, can be monitored. When the solute is desorbed, a decrease in temperature takes place. The main advantage of this detector is that it has a very low dead-volume and all solutes are qualitatively visualized. This can be important if it is necessary to determine all the constituents present in a sample. It is easy to use, is inexpensive, can be purchased as a modular unit, and has minimum drift. However, it is less suitable for quantitative analysis than other techniques because there is great variation in the heat of adsorption and the detector requires absolutely constant ambient temperature. Also, the flow rate must be constant and surge-free. When a gradient elution is used, a shift in the baseline will

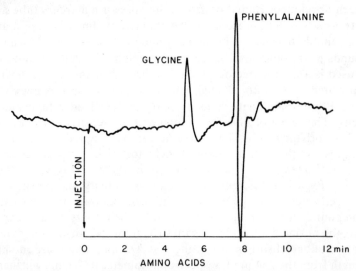

Fig. 2–14. Use of microadsorption detector. *Instrument*: Varian LCS 1000. *Eluent*: 0.2 N sodium citrate, pH 3.25. *Flow rate*: 25 ml/hr. *Column*: 30 cm × 0.6; packed with UR^{-30} resin (27 cm-20μm). *Sample*: 100 μg glycine; 100 μg phenylalanine. (Contributed by D. Gere of Varian Aerograph.)

occur when the solvent composition changes. A sample of the type of chromatogram of amino acids obtained with a microadsorption detector by Gere (1970), is shown in Fig. 2–14. The peak shape of the microadsorption detector has been named the "differential peak shape" because of its similarity to the derivative of the gaussian curve and because it depends on the differential in heat input and heat loss. Although details of the peak shape vary from one sample–carrier–adsorbent combination to another, parameters such as relative height and width of the positive and negative components of the peak depend on the adsorption isotherm and heat transfer characteristics of the particular combination. Typical applications for this detector are carbohydrates, essential oils, lipids, triglycerides, nucleotides, and amino acids (Fig. 2–15).

D. Electrolytic Conductivity Detector

Electrolytic conductivity has also been utilized as a detector in liquid chromatography with aqueous mobile phases and ionic solutes. The specific conductivity of the effluent stream is monitored continuously and the solute in the solution detected by a change in conductivity. Although it is one of the simplest and most dependable of the monitoring techniques presently in use, cell parameters and measuring circuitry have limited its use. However, they are being improved to permit use with both aqueous and nonaqueous solvents. In this technique a small volume flow through a cell is used and variations in the conductivity of the effluent due to the presence of eluted solutes are continuously monitored and recorded. Response is linear with concentration over a wide range, but gradients cause a shift in the baseline. This method of detection is especially applicable

Fig. 2–15. Application of Microadsorption detectors. *Instrument*: Varian LCS 1000. *Sample*: (1) carbohydrate, 10 μg fructose; (2) triglyceride, 100 μg tribehenin; (3) lipid, 0.91 mg triolein; (4) amino acid, 450 μg proline; (5) nucleotide, 50 μg guanosine 5′ monophosphate. *Sensitivity*: (1) × 32; (2) × 32; (3) × 32; (4) × 1; (5) × 1. *Adsorbant*: (1) Amb. CG-400; (2) silicic acid; (3) silicic acid; (4) Sephadex G-10; (5) DE-32. *Carrier*: (1) 0.1 *M* sodium borate; (2) benzene; (3) benzene; (4) water; (5) 0.1 *M* sodium acetate + 0.22 *M* acetic acid. *Flow rate*: (1) 25 ml/hr; (2) 45 ml/hr; (3) 45 ml/hr; (4) 14.2 cm³/hr; (5) 45 cm³/hr. (Contributed by Varian Aerograph.)

to gel filtration chromatography, but the simplicity of operation, ease of cleaning, and freedom from maintainance and calibration make it advantageous for use in other types of chromatography. Chromatronix and Laboratory Data Control now market a conductivity detector with a microcell for use with a variety of liquid chromatography systems. Conductivities are listed in Table 2–5.

E. FLAME IONIZATION DETECTOR

The flame ionization detector, which is so popular in gas chromatography, can also be used in liquid chromatography. Its main limitation in liquid chromatography, however, is that the sample must be volatile, and the mobile phase a volatile, organic solvent. Also, the solute is destroyed in the detection process and therefore the fractions cannot be collected. However, it is highly sensitive, has a wide dynamic and quantitative

TABLE 2–5

CONDUCTIVITIES[a]

K_{sp} Specific conductance μmho/cm	Approximate conductivities $K_{sp} = K \times$ cell constant	K Conductance with MCC-75 cell μmho
0.01 —		
0.1 —	— Ultrapure water —	— 0.001
1 —	— 0.1 ppm NaCl —	— 0.01
	— Good quality distilled water —	
	10^{-5} M NaCl —	
10 —	— 10^{-4} M NaCl —	— 0.1
100 —	— Excellent quality raw water —	— 1
	10^{-3} M NaCl	
1000 —	— 10^{-2} M NaCl —	— 10
10K —	— 10^{-1} M NaCl —	— 100
100K —	5% NaCl	— 1000

[a]Contributed by Chromatronix, Inc.

response, and can be used with gradient elutions. It has been used mainly on lipids, proteins, and polymers. In order for a flame ionization detector to work, the eluent must be separated from the sample before combustion can take place. In the commercial flame ionization detectors now available, all or a portion of the column effluent is applied onto a small, moving chain or wire and the solvent is evaporated. Then the sample residue is volatilized and its vapors are driven by an inert gas stream into the flame ionization detector. Detector noise is not seriously affected by minor flow or temperature variations. Although the microadsorption detector can be used for lipid separations, the flame ionization detector is the detector of choice for this class of compounds. In the future it may prove to be as useful as it has been in gas chromatography.

F. Other Detectors

1. Polarography

Two new microdetectors have been designed based on the use of polarography. One constructed by Koen *et al.* (1970) used a dropping mercury electrode. The other, designed by Joynes and Maggs (1970), uses a commercially available polarographic electrode. Polarography is an electrochemical analysis method. The mobile phase must have an aqueous mixture of acids and salts. Thus this technique is limited to aqueous effluents. It is, however, one of the most useful detectors in performing inorganic analyses. Kemula (1952) first used liquid chromatography in conjunction with polarography for the analysis of metal ions. He called the method chromatopolarography. It is only recently, however, that a polarographic detector for high speed liquid chromatography has been designed. Normally, in polarography, a solution is electrolyzed by using a small polarizable, dropping electrode and a large nonpolarizable electrode. The current increases stepwise as a function of the voltage applied to the electrode when the solution contains solutes that can be reduced or oxidized at the dropping mercury electrode. The halfwave potential is the potential at which a current step appears and is characteristic of that type of compound reacting at the dropping mercury electrode. The diffusion of the electrochemically active substance to the dropping mercury electrode controls the maximum current of a step which is proportional to the concentration. There are two major disadvantages of using a dropping mercury electrode for continuously monitoring the effluent. One is the need for damping the noise associated with the dropping mercury since the noise frequency is a function of the drop rate. The second is the need to remove oxygen from the eluent since it interferes with the detection of other solutes. For these reasons, Joynes

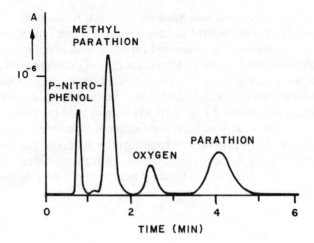

Fig. 2–16.   Use of polarographic detector. *Instrument*: noncommercial. *Sample*: chromatogram of a technical insecticide mixture. Injected 10 μl of a 0.4% (v/v) solution of Folidol in the eluent. *Flow rate*: 60 ml/hr. Time constant of the single RC circuits of the damping network, $\frac{1}{2}$ sec, [Contributed by Koen *et al.* (1970).]

and Maggs (1970) did not use a dropping mercury electrode, but instead their polarographic detector utilized a carbon-impregnated silicone rubber membrane as the electrode. The advantages that this electrode has over the dropping mercury electrode are that it has low residual or standing currents, it is less prone to oxide films, it is simple and robust, it has low noise level and therefore does not require damping, and the interference of oxygen in the effluent is much lower. This electrode is commercially available and suitable for oxidation and reduction of organic and inorganic substances. An example of the use of polarography as a detector for the separation of insecticides is shown in Fig. 2–16 (Koen *et al.*, 1970), and the precision of quantitative measurements in Table 2–6.

2. Radioactivity Monitor

Although a detector to measure the radioactivity of the effluent is not available commercially for liquid chromatography, the advantages of such a device would be good quantitation, wide dynamic range, utility with gradient elution, and low minimum detectable sample size. Since the solubility properties of calcium fluoride approach those of glass for most solvents used in liquid chromatography, calcium fluoride scintillators give promise of extending this monitoring technique to many solvent systems which

TABLE 2-6

PRECISION OF QUANTITATIVE MEASUREMENTS USING A POLAROGRAPHIC DETECTOR[a]

Syringe capacity (μl)	Sample volume (μl)	Sample concentration (moles/liter)			Sample amount (ng)			Relative standard deviation (%) of the peak area			Number of measurements
		p-Nitro phenol	Methyl parathion	Parathion	p-Nitro phenol ($M = 139$)	Methyl parathion ($M = 263$)	Parathion ($M = 291$)	p-Nitro phenol	Methyl parathion	Parathion	
25	25.0	5.0×10^{-4}	1.33×10^{-3}	3.09×10^{-3}	1750	8750	22.500	4.5	6	4	4
10	5.0	5.0×10^{-4}	1.33×10^{-3}	3.09×10^{-3}	350	1.750	4.500	1.7	1.3	2	6
10	1.0	5.0×10^{-5}	1.33×10^{-3}	3.09×10^{-3}	70	350	900	2.8	2.4	2.6	6
10	5.0	2.16×10^{-5}	6.08×10^{-5}	1.37×10^{-4}	15	80	200	3	3.3	3.8	4
10	1.0	2.16×10^{-5}	6.08×10^{-5}	1.37×10^{-4}	3	16	40	12	5.7	5	7

[a]Contributed by Koen et al., (1970).

previously could not be used. In this technique, a column effluent would pass through a cell or tube packed with a scintillator which has a suitable efficiency for ^{14}C β-rays. A photomultiplier detects the light pulses generated, which are counted or converted to a voltage level with a ratemeter.

3. Fluorescence

Fluorescence is the absorption of UV radiation and the reradiation of light at a longer wavelength. The main advantages of fluorescence monitoring are its specificity and the fact that small quantities of strongly fluorescent solutes can be detected. Although fluorescence is possible with gradient elution chromatography, it is absolutely necessary that the solvents used are transparent, both to the exciting UV and the fluorescence wavelength. A Fluoromonitor is being marketed by Laboratory Data Control which has dual flow cell volume of 10 μl and good photometer stability. Excitation is standard at 360 nm but other wavelengths are also available. It is sensitive to concentrations as small as 1 ppb and can detect 1×10^{-11} gm quinine sulfate. This detector can be used alone or in conjunction with a refractive index or UV monitor.

4. Dielectric Constant Monitor

When the polarity of solute and solvent differ widely, the dielectric constant monitors can be used to best advantage because dielectric constants or capacitance monitors respond to the polarity and to the polarizability of the materials between the plates. The dielectric constant is the electrical analog of refractive index. The monitors therefore exhibit many of the same advantages and disadvantages. For example, as with most physical detectors, they are nondestructive to the sample, provide quick response, and contribute little to extracolumn-band broadening. The response is linear with concentration; thus there can be direct quantitation with area analysis. However, it is not as sensitive as the UV monitor. The main use of the dielectric monitor, as yet, is with gel filtration chromatography.

a. Buffer Storage. Karmen *et al.* (1970) have devised a system for increasing the productivity of quantitative liquid chromatographic systems. Increased detection sensitivity, as well as a more effective utilization of detection systems, can result if the separation and quantitation functions are performed at different times. Therefore they propose using an off-line approach in which separation and detection processes are performed independently, rather than the practice of measuring on-line the concentration in the column effluent while the separation is occurring. Many separations can occur simultaneously and each separation preserved in a simple buffer storage for subsequent rapid quantitation.

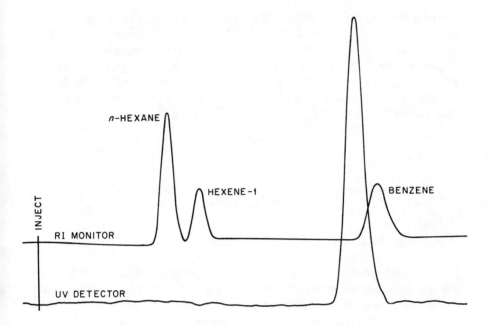

Fig. 2–17. Use of combination of ultraviolet and refractive index detectors. *Instrument*: Waters ALC 202 liquid chromatograph. *Column*: 4 ft × 2.3 mm; packed with Porasil T. *Eluent*: Perfluoro (methyl cyclohexane). *Flow rate*: 0.5 ml/min. *Pressure*: 970 psig. *Sample*: n-hexane, hexene-l, and benzene. (Contributed by Waters Associates, Inc.)

b. Multiple Detector Systems. Although the RI detector is a universal detector, its sensitivity (1 ppm) is much lower than that of the UV detector (0.06 ppm). However, the latter is sensitive only to materials having UV absorbance at the set wavelength and is completely insensitive to those compounds that do not absorb. An example of using dual detectors (UV and RI) to separate organic compounds is shown in Fig. 2–17. The heat of adsorption and UV detectors are also being used in combination. It is also possible to obtain systems with dual heat of adsorption and RI detectors. Since detectors are available as modular units, detectors can be combined to provide most the information and best results for the problem at hand. When dual detectors are used, usually dual pen recorders are used. Laboratory Data Control, Inc. sells a dual UV monitor using UV absorption at 254 and 280 nm with a recorder which records the differential absorbance at the two different wavelengths. Together with a dual pen recorder, much more useful information is obtained than with either recorder alone.

It can be seen that there is still no true universal detector for high pressure liquid chromatography. Because of the need to find better detectors to monitor the liquid chromatographic methods, research in this area must continue. The greatest need at present is for a widely applicable detector that can accommodate solvent gradient with no loss in sensitivity.

More and better detectors must be developed for use in preparative high pressure liquid chromatography since there is a growing need for the preparation of pure fractions. It is proposed that in the future, multiple detectors may be used either in series or by stream splitting prior to detection.

IV. Columns

A. COLUMN PACKINGS

The performance of a chromatographic system is determined by the separation achieved on the column. More literature has been written about column-packings and design than about all the other components of the system. Excellent reviews on columns for liquid chromatography were written by Kirkland (1971b) and Horvath (1972b).

The stationary phase should be thermally stable and chemically inert to both the solvent of the mobile phase and the solutes. Based on the classical work of Cohn (1949, 1950, 1955), Cohn and co-workers (1951, 1961), and Volkin *et al.* (1951) who developed ion exchangers as column packings for the separation of bases, nucleosides, and nucleotides, Anderson *et al.* (1963) prepared columns 0.9 × 150 cm using Dowex 1-8x resin. Although the separations were good, because of the low pressure and slow flow rate, the complete elution schedule was 28 hours. Moreover, it was found that there were other difficulties in using conventional ion exchange resins. A strong gradient is required to separate the mono-, di-, and triphosphate nucleotides because of the large differences in the equilibrium constants; thus there is swelling and shrinking of the column packing. This may cause changes in column performance and permeability. Low molecular weight contaminants may cause column bleeding, baseline drift, and interference when collecting the eluted compound.

Halász and Horvath (1964a,b) experimented with microglass beads coated with a porous thin layer as a column packing in gas chromatography and in 1967 Horvath *et al.* applied this type of packing to a high pressure liquid chromatography system. A pellicular ion exchange resin was developed which consisted of microglass beads (~ 50 μm diameter) coated with an ion exchange resin. Directions for the preparation of these stable, efficient resins are described by Horvath in his review "Ion Exchangers" (1972b). These pellicular anion exchange resins permitted the rapid and sensitive analysis of nucleotides (Horvath *et al.*, 1967). On the cation exchangers, fast separation of the nucleosides and bases was obtained (Horvath and Lipsky, 1969b).

It was found that the speed of separation depends to a large extent on the packing of the column. The speed can be increased by using column packing that is uniform in dimension, of solid core, and with a thin porous layer adsorbant. Kirkland (1969b) reported on the development of packings made with a controlled surface porosity support for liquid–liquid chromatography and investigated the critical parameters. These include the liquid film thickness, the types of support, the support surface porosity, the internal diameter of the column, and the reproducibility of the column preparation. The support used consisted of siliceous particles with a porous surface of controlled thickness and pore size. The practical advantage of this type of support is that it can be packed dry. Later, reproducible, high efficiency column packings were prepared with a chemically bonded organic stationary phase (Kirkland and DeStefano, 1970). A variety of functional groups are available; thus there is a widely diverse selection of chromatographic columns in which the packings are nonextractable and thermally and hydrolytically stable. High carrier velocities can be used and good column efficiency maintained. Because the packings are stable and there is minimal column bleeding, the column has a long life. These column materials were prepared by reacting reagents with the surface porous shell on the support. The agents were then polymerized to give the desired silica coating. Controlled surface porosity anion and cation exchange packings were also prepared (Kirkland, 1970), and it was found that rapid separation of nucleotides and bases could be obtained. The cation exchange medium is a fluoropolymer containing very strong sulfonic acid groups. The anion exchanger consists of methacrylate polymers or polystyrene with tetrammonium groups as exchange sites. These packings have high mechanical stability and low swelling characteristics, thereby permitting inlet pressures up to 5000 psi without damage to the column. Henry and Schmit (1970) used this type of column to successfully separate carboxylic acids.

Little *et al* (1970a) also compared the performance of porous and solid core porous adsorbants. They coated spherical particles of a solid glass core with a thin porous coating of silica, and also coated solid spheres with a double layer of porous silica. Other packings were prepared of spherical, porous, silica beads of different surface areas and still another of irregularly shaped porous silica of high surface area. It was found that by using a solid core support, the speed of analysis was greatly increased. Columns of these compounds gave substantially less peak broadening and because of better mass transfer properties, improved separations were obtained. Solid core packings are generally highly efficient and provide narrower peaks, thus allowing detection of minute quantities of a solute in the effluent. The requirements for good column packing materials for high pressure liquid chromatography therefore include large surface areas, a thin layer of adsorbant, uniformly dispersed, and open-structured surfaces readily accessible to the mobile phase. Also, it must be possible to pack the material uniformly. The packing must be stable, not easily compressed by high pressures, nor easily disturbed by high velocities. Thus it can be concluded that small particles of uniform spherical shape, which have a solid core and porous layer, give the best performance in high pressure liquid chromatographic systems. A table of some of the packings commercially available are listed in Table 2–7. An example of the difference in separation that can be caused by the column packing is shown in Fig. 2–18. Using Corasil I, a spherical glass particle with a thin porous silica layer, six insecticides were separated in 2 minutes. When Corasil II, which has a

TABLE 2–7

TABLE OF COLUMN PACKINGS

Support	Type	Supplier	Use
Corasil	Porous layer bead	Waters	LSC, LLC
LC Durapak	Chemically bonded Porasil	Waters	LSC, LLC
Durapak/Corasil	Chemically bonded Porasil PLB	Waters	LSC, GPC
Aquapak	Rigid crosslinked polystyrene	Waters	LSC, GPC
Poragel	Crosslinked polystyrene gel bead	Waters	LSC, GPC
Porasil	Rigid, porous silica	Waters	LSC, LLC, GPC
EM gel-Type SI	Crushed porous silica bead	Waters	LSC, LLC, GPC
EM gel-Type OR	Semirigid crosslinked polystyrene acetate	Waters	LSC, LLC, GPC
Corning controlled porosity glass	Etched glass beads	Corning	GPC
Zipax	Porous layer bead	du Pont	LLC
Permaphase ODS	Chemically bonded PLB	du Pont	LLC
Sil-X	Surface modified silica	Nester/Faust[a]	LSC
Pellionex	Pellicular ion exchange	Northgate Lab.[b]	Ion exchange
Pellosil	Pellicular silica	Northgate Lab.	LSC or support in LLC
Pellumina	Pellicular alumina	Northgate Lab.	LLC
Pellidon	Pellicular porous polyamide	Northgate Lab.	LSC

[a]Perkin-Elmer.
[b]Reeve Angel.

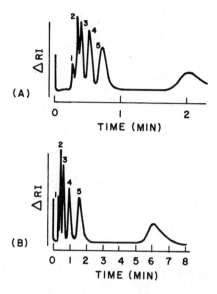

Fig. 2–18. Effect of packing on separation of insecticides. *Instrument*: Waters ALC 202. *Column*: 20 cm × 2.3 mm i.d., packed with (A) Corasil I and (B) Corasil II. *Eluent*: hexane. *Flow rate*: 3.0 ml/min. *Sample*: (1) aldrin impurity; (2) aldrin; (3) p,p'-DDT; (4) DDT; (5) lindane; (6) endrin. [Contributed by Little *et al.* (1970a).]

double layer of the porous silica, was used, the separation took longer and the components that came off early in the chromatogram were not as well resolved. Four different packings were compared by Little *et al.* (1970b) in the separation of 2,3-xylidine and 2,6-xylidine and the speed of analysis compared in Table 2–8. In Table 2–9 the comparisons of column efficiencies using 2,3-xylidine as a sample are tabulated.

B. COLUMN DESIGN

It has been found by Snyder (1969b) that the efficiency of a column depends on the bed structure. The bed structure, in turn, is determined by several factors such as the packing material, average particle diameter, the technique used to fill the column, and the geometry of the column. In the development of high pressure liquid chromatography, the design of the column has evolved with the instrumentation. Columns for conventional column chromatography are usually glass; in gas chromatography, however, copper, aluminum, glass, or stainless steel have been used. In most cases,

TABLE 2–8

EFFECT OF DIFFERENT PACKINGS ON SPEED OF ANALYSIS[a]

A. *Instrument:* Waters ALC–100.   *Solvent:* 10% isopropanol in hexane.
Flow rate: 1.55 ml/min.   *Sample:* 2,3-xylidine; 2,6-xylidine.

Packing	R^b	t^c (sec)	R/t
Corasil I	0.58	41	1.41×10^{-2}
Corasil II	0.84	47	1.79×10^{-2}
Porasil 400	1.29	120	1.07×10^{-2}
Porasil T	1.92	149	1.29×10^{-2}

B. *Instrument*: Waters ALC-100.   *Solvent*: 1% isopropanol in hexane.
Flow rate: 1.55 ml/min.   *Sample*: 2,3-xylidine; aniline.

Corasil I	0.82	56	1.46×10^{-2}
Corasil II	1.15	89	1.29×10^{-2}
Porasil 400		Peaks skewed	
Porasil 1500	0.69	102	0.68×10^{-2}
Porasil T	1.91	491	0.39×10^{-2}

[a]Contributed by Little *et al.*, (1970).
[b]R is the resolution which is expressed in terms of k', α, and N where $k' =$ capacity factor of the column to retain samples; $\alpha =$ separation factor of the column; and $N =$ efficiency of the column (theoretical plate number).
[c]t is the time in seconds between the two peaks.

TABLE 2–9

COMPARISON OF COLUMN EFFICIENCIES[a,b]

Packings	Plates/ft	k'^c
Corasil I	1050	0.18
Corasil II	1200	0.36
Porasil 400	228	1.24
Porasil T	703	1.19

Packings	Sample	k'^c	Plates/ft	Width (ml)
Corasil II	Aniline	0.49	1235	0.131
Porasil 400	2,6-xylidine	0.58	554	0.341
Porasil T	2,6-xylidine	0.43	885	0.314

[a]Contributed by Little *et al.*, (1970a).
[b]*Solvent*: 10% isopropanol in hexane.   *Linear velocity*: 0.3 cm/sec.   *Sample*: 2,3-xylidine.
[c]k' is the capacity factor of the column to retain samples.

$$k'_1 = \frac{V_1 - V_0}{V_0}$$

where V_1 is the elution volume of a peak, and V_0 is the void volume.

stainless steel tubing has been found to be most practical in both gas and high pressure liquid chromatography. Although straight columns are generally more efficient, they are more cumbersome, and small-bore coiled columns have been found to be preferable. It was found that better efficiency was obtained with narrow bore columns. Anderson *et al.* (1963) used a 9 mm i.d. column but Horvath *et al.* (1967) found that 1 mm i.d. stainless steel tubes, which were dry packed, were very efficient. Felton (1969) investigated columns of 1.5, 2, and 3 mm i.d. and found that the smaller bore columns were advantageous for good resolution, as did Bombaugh *et al.* (1970). However, R. P. W. Scott *et al.* (1967) and Snyder (1969b) used a column approximately 4 mm i.d. and obtained good results. Kirkland (1969b) used three 1000 × 2.1 mm straight columns in series with little loss in total efficiency. It has also been found that the type of tubing used in the columns greatly affects their efficiency and stainless steel tubing is the material of choice by the majority of researchers. Recently, DeStefano and Beachhell (1970) used columns with large diameters (7.94 and 10.0 mm) packed with controlled surface porosity support and obtained excellent results by monitoring only the central cut of the effluent. This procedure was described by Knox and Parcher (1969). The use of the large diameter column will be of great use in preparative chromatography, since the small bore column necessitates the use of small injection volumes.

V. Recorders, Fraction Collectors, and Integrators

To obtain a permanent record of results, a strip chart recorder is generally used. With high pressure liquid chromatography, a potentiometric type recorder with a 10-mV, 1-second full scale response is recommended, but a 1 mV recorder or one with variable response may be used. The recorder used, however, is dependent on the type of detector. The characteristics which are desirable are that it should have a linear range, zero shifting, and pen speed with full scale response of 1 second or less. Recorders are available separately or can be purchased with the instrument. It is possible to obtain recorders with single or dual channel recording and it is advisable to have variable chart speeds. An event marker is helpful if fractions are to be collected.

Fractions can be collected manually or automatically by drop or time. There are many automatic fraction collectors manufactured and almost any of them can be connected to a high pressure liquid chromatographic system. If liquid chromatography is to be used in conjunction with isotopic labeling, the most convenient setup is to be able to collect the fractions directly in scintillation vials because it eliminates one step in the transferring process.

Integrators can be purchased as part of a liquid chromatographic system or separately from the instrument company or any one of a number of companies that market integrators for gas and liquid chromatographs as well as for other laboratory instruments. Many times an integrator is not necessary at first, but as the liquid chromatograph is utilized more for many research projects, an integrator is no longer a luxury but a time-saving necessity. Also, since more consistently accurate results are obtained with an electronic integrator than with any manual method of measuring area, for precise quantitative results an integrator should be acquired. In 1969 Gill wrote a review of chromatographic automation for gas chromatographs and Hegedrus and Petersen (1971) discussed thoroughly the integration of chromatographic signals by digital computers. The use of digital integrators with on-line computers was discussed by Chilcote et al. (1971) and the possibility of interfacing chromatographs with computer systems by Lyons (1972). A good review of the trends in automation for chromatographic systems was written by Anderson (1972). The use of integrators with high pressure liquid chromatographs is discussed more fully in Chapter 5.

CHAPTER 3

‖

EXPERIMENTAL METHODS

High pressure liquid chromatography is becoming increasingly popular as an analytical tool because of its versatility. As in gas chromatography, the stationary phase, the flow rate, and the temperature can be varied; liquid chromatography, however, has the added advantage of being able to change many of the variables of the mobile phase. For example, the mobile phase can be either an aqueous or organic solvent. If aqueous solutions are used, pH, ionic concentration, and anions or cations present can be varied. An isocratic elution mode, stepwise or a gradient elution, can be utilized. If a gradient is preferred, it can be linear, concave, or convex and the slope of the linear concentration gradients can be varied.

In order to obtain the best possible separations, the operating conditions for each group of compounds must be investigated and optimized. Although some of these conditions can be predicted, others must be determined experimentally. A new column, even though it is packed with the same kind of resin as previously used, may require new conditions to achieve the best resolution. Guidelines for determining the conditions to adjust the operating conditions have been outlined by Cohn (1949), Horvath *et al.* (1967), Uziel *et al.* (1968), and Burtis (1970a). It is best to change one variable at a time; however, it may be necessary eventually to adjust more than one operating parameter to achieve the best results possible.

In summarizing guidelines for optimization of operating conditions for high pressure ion exchange liquid chromatography, Horvath and Lipsky (1969b) recommended the determination of the following variables:

51

1. The resin properties (particle size, structure of polymer, shell thickness, etc.)
2. Chemical composition of eluents
3. Concentration of eluents
4. pH of eluents
5. Column temperature
6. Length and diameter of column
7. Flow rate
8. Elution modes

I. Standardization of Conditions

In order to be able to check the reproducibility of the results and the stability of instrument and column conditions, a standard solution should be analyzed routinely. The standard should be typical of the type of compounds being separated. For example, in the analysis of nucleotides, Brown (1970) analyzed a standard solution of adenine and guanine nucleotides at

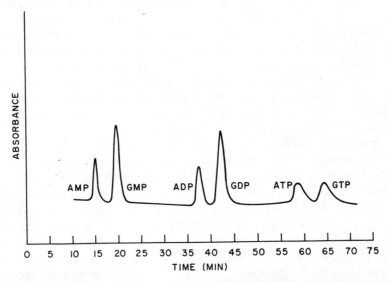

Fig. 3–1. Nucleotide standards. *Instrument*: Varian LCS 1000. *Column*: 1 mm i.d. × 3 m; packed with pellicular anion exchanger. *Eluents*: 0.015 M KH$_2$PO$_4$ and 0.25 M KH$_2$PO$_4$ in 2.2 M KCl. *UV output*: 0.08 o.d. *Flow rate*: 12 ml/hr. *Samples*: 3 μl of a mixture of ∼0.5 mM AMP, ADP, and ATP; 3μl of a mixture of ∼0.5 mM GMP, GDP, and GTP. [Contributed by Brown (1970).]

TABLE 3–1

EFFECT OF STORAGE ON CONCENTRATION (MILLIMOLAR) OF STANDARD NUCLEOTIDE
SOLUTIONS AS DETERMINED BY VARIAN AEROGRAPH LCS 1000[a]

Sample no. [b]	Time of storage	AMP	ADP	ATP	GMP	GDP	GTP	ATP/ ADP
SR-0	0	0.32	0.49	0.49	0.32	0.33	0.29	1.0
SR-1	1 week	0.25	0.41	0.41	0.28	0.27	0.27	1.0
SR-2	1 month	0.32	0.42	0.40	0.32	0.25	0.26	0.9
SR-3	2 months	0.39	0.31	0.38	0.29	0.18	0.24	1.2
SF-0	0	0.32	0.49	0.49	0.32	0.33	0.29	1.0
SF-1	1 week	0.33	0.45	0.49	0.31	0.32	0.28	1.0
SF-2	1 month	0.33	0.51	0.52	0.34	0.35	0.28	1.0
SF-3	2 months	0.30	0.44	0.45	0.32	0.30	0.28	1.0
SD-0	0	0.32	0.49	0.49	0.32	0.33	0.29	1.0
SD-1	1 week	0.24	0.41	0.48	0.30	0.28	0.24	1.0
SD-2	1 month	0.31	0.44	0.46	0.31	0.31	0.26	1.0
SD-3	2 months	0.28	0.44	0.46	0.29	0.31	0.28	1.0

[a] Brown (1971b).
[b] The "R" samples are those stored in the refrigerator; the "F" samples in the freezing
unit of the refrigerator at $-20°C$; and the "D" samples in the deep freeze at $-80°C$.

least once a week, if not more often, to make sure that the retention times
did not change, the resolution remained the same, and the baseline was
stable (Fig. 3–1). It has been found by Brown (1971b) that nucleotides, when
stored frozen as neutral solutions, are stable (Table 3–1) and there is little
deterioration; thus one or two of the nucleotides in a solution make good
reference materials for other nucleotides. In working with other groups of
compounds, reference compounds should be found that are stable, separate
well under the operating conditions used, and run the gamut of retention
times for the compounds being analyzed. It is of critical importance that
controls be run in any experiment, and most especially when high pressure
liquid chromatography is being used as an analytical tool because of its great
sensitivity. Since the development of this instrumentation is fairly recent,
there is not a large accumulation of reference literature on the optimum
conditions for separating the vast numbers of compounds possible as there
is in gas chromatography today. However, the Institute of Petroleum has
published a comprehensive index to liquid chromatography, "Gas and
Liquid Chromatography Abstracts," which covers the field of high speed
liquid chromatography from 1966 to date. It is possible to obtain references
to specific compounds in *Chemical Abstracts* and some of the journals and
monographs have published reviews on specific aspects of high pressure
liquid chromatography. For example, an excellent review, "High Perform-

ance Ion Exchange Chromatography with Narrow Bore Columns: Rapid Analysis of Nucleic Acid Constituents at the Subnanomole Level" by C. G. Horvath, will soon be published in "Methods of Biochemical Analysis" (1972a). A book edited by J. J. Kirkland (1971a), "The Modern Practice of Liquid Chromatography," which was the result of a course given on high pressure liquid chromatography, presents both theoretical and practical information on this technique. Also the manuals put out by the various instrument companies on liquid chromatography are excellent sources of reference material.

For each experiment, carefully designed controls must be set up and run to assure valid results. The extraction procedure, sampling techniques, and storage methods must be checked thoroughly and all solvents examined to make sure that they, or impurities present in them, do not interfere with the analysis. For example, when experiments are being run on cell extracts, a backlog of samples may result. In order to check the stability of these extracts, Brown (1971b) investigated the effect of storage at different temperatures. The results are shown in Table 3–2.

TABLE 3–2

EFFECT OF STORAGE ON THE CONCENTRATION OF ADENINE AND GUANINE NUCLEOTIDES OF HUMAN ERYTHROCYTES (MILLIMOLES/LITER OF PACKED RED BLOOD CELLS) AS DETERMINED BY THE VARIAN AEROGRAPH LCS 1000[a]

Sample	[b] Time of storage	AMP	ADP	ATP	GDP	GTP	AMP, ADP, and ATP	ATP/ ADP
CR-0	0	0.06	0.17	0.91	0.02	0.06	1.14	5.5
CR-1	1 week	0.02	0.19	0.97	0.03	0.06	1.18	5.1
CR-2	1 month	0.02	0.23	0.97	0.02	0.05	1.22	4.3
CR-3	2 months	0.10	0.44	1.78	0.03	0.11	2.32	4.0
CF-0	0	0.02	0.17	0.90	0.02	0.04	1.09	5.4
CF-1	1 week	—	0.16	0.92	0.01	0.04	1.07	5.4
CF-2	1 month	—	0.19	0.95	0.03	0.05	1.14	5.0
CF-3	2 months	—	0.19	0.95	0.04	0.08	1.16	5.0
CD-0	0	0.02	0.17	0.92	0.03	0.08	1.11	5.4
CD-1	1 week	0.02	0.17	0.92	0.03	0.07	1.11	5.4
CD-2	1 month	—	0.19	0.92	0.02	0.05	1.11	5.0
CD-3	2 months	—	0.19	0.92	0.02	0.06	1.11	5.0

[a]Brown (1971b).

[b]The "R" samples are those stored in the refrigerator; the "F" samples in the freezing unit of the refrigerator at $-20°C$; and the "D" samples in the deep freeze at $-80°C$.

II. Operating Parameters

A. TEMPERATURE

The optimum column temperature must be determined for each class of compounds and for the operating conditions involved. The effect of temperature in liquid chromatography has been studied by a number of workers. Le Rosen and Rivet (1948) and Lie Tien (1953) investigated temperature effects on solute retention in liquid chromatography. Martire and Locke (1967) studied this effect in liquid–liquid systems and Dybeznski (1967) and Horvath *et al.* (1967) examined temperature dependence in ion exchange chromatography. Hesse and Englehardt (1966) investigated the possibility of temperature programming and Obrink *et al.* (1967) and Leach and O'Shea (1965) found that temperature can affect column performance in gel permeation chromatography. Maggs (1968) investigated the effect of temperature on a liquid–solid chromatographic system and found that an increase in temperature caused greater activity of the exposed porous sites and increased the polarity of a polar, stationary phase, thus increasing the retention volume. Maggs also found that between ambient temperatures and 80°C,

TABLE 3-3

EFFECT OF COLUMN TEMPERATURE[a]

Chromatographic performance factor	Reduced temperature	Constant temperature	Increased temperature
Retention volume	Increased	Stable	Decreased
Sample stability	Increased		Decreased
Stationary phase thermal stability	Increased	Stable	Decreased
Solubility of Stationary phase in mobile phase	Decreased	Stable	Increased
Analysis time (all factors constant except column temperature)	Increased	Stable	Decreased
Viscosity and pressure drop	Increased		Decreased
Flow rate (at constant pressure)	Decreased	Stable	Increased
Resolution due to retention volume	Increased		Decreased
Resolution due to peak spreading	Decreased		Increased

[a]Contributed by Waters Associates (1970).

the higher the operating temperature, the greater the resolution, due mainly to the distribution coefficients and retention ratios of the solutes with temperature. However, the long time needed to achieve equilibrium between the mobile and stationary phase when the temperature is changed is a definite disadvantage in the use of temperature programming in some liquid–solid chromatographic systems. R. P. W. Scott and Lawrence (1969) also found pressure programming or gradient elution to be more advantageous than temperature programming for this reason. Snyder (1969a) compared solvent, flow, and temperature programming in liquid chromatography. From his work in liquid–solid chromatography, he found that only elution gradient is useful for wide-range samples and that flow and temperature programming appear rather limited in improving the performance. Although it has been found in liquid chromatography that the selection of the mobile and station-

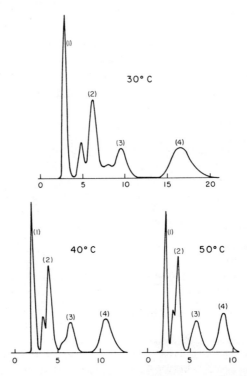

Fig. 3–2. Effect of temperature on the separation of 2'- and 3'-monophosphate nucleotides. *Instrument*: noncommercial. *Column*: pellicular strong cation exchange resin, 1 mm i.d. × 152 cm *Eluent*: 0.01 $MNH_4H_2PO_4$, pH 2.5. *Flow rates*: 14.8 ml/hr (30°C); 18.7 ml/hr (40°C); 19.3 ml/hr (50°C). *Sample*: 750 μM of each compound: (1) uracils, (2) guanines, (3) adenines, (4) cytosines. [Contributed by Horvath (1972b).]

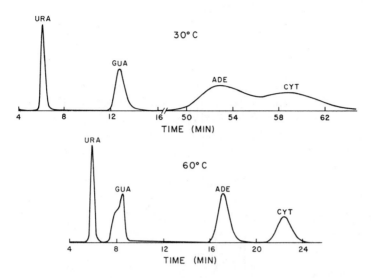

Fig. 3–3. Effect of temperature on the separation of bases. *Instrument*: Pickar (now Varian) LCS 1000. *Column*: 1 mm i.d. × 300.7 cm; packed with pellicular strong cation exchange resin. *Eluent*: 0.02 *M* KH₂PO₄, pH 5.2. *Flow rate*: 14.8 ml/hr. [Contributed by Horvath and Lipsky (1969c).]

ary phases are the most powerful variables in optimizing separations, column temperature can have some effect on chromatographic performance. The general effects of column temperature control have been well summarized by Waters Associates (1970) (Table 3–3). Horvath *et al.* (1967) found that increasing the column temperature to 70°C aided in more rapid separations of the nucleotides and noted no decomposition of these relatively thermally labile compounds at this temperature. Brown (1970) found that in the separation of nucleotides in cell extracts that 75°C was optimal, especially for peak sharpness of the triphosphate nucleotides. However, although it was also found that temperature and pH were not as important as ionic strength and slope of the linear concentration gradient in attaining good resolution of nucleotides, the effect of increasing the temperature in the separation of 2′- and 3′-monophosphate nucleotides is illustrated in Fig. 3–2. The effect of temperature on the retention time of the pyrimidine and purine bases is different and chromatograms at 30° and 60°C of uracil, guanine, adenine, and cytosine are shown in Fig. 3–3. At 60°C the retention times of the adenine and guanine are greatly reduced and the peak shapes markedly sharpened. Horvath and Lipsky (1969b) found that the optimum temperature is different for bases and nucleosides. For nucleosides, although the retention

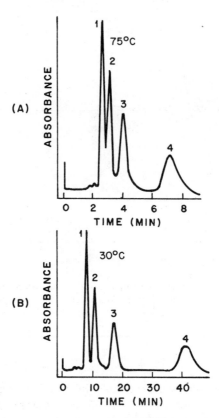

Fig. 3–4. Effect of temperature on liquid–liquid chromatography. *Instrument*: Waters ALC-202 liquid chromatograh. *Column*: 4ft; packed with 1.3% squalane on Corasil I. *Solvent*: 1/1 water/methanol. *Pressure*: 1000 psi. *Samples*: (1) anthraquinone; (2) 2-methylanthraquinone; (3) 2-ethylanthraquinone; (4) 2-tert-butylanthraquinone. [Contributed by Waters Associates, Inc. (1970).]

times are longer, higher resolution was found at or below 40°C. An example of the effect of an increase in temperature on retention time in a liquid–liquid separation is illustrated in Fig. 3–4.

By reducing column temperature, retention times are generally increased. This can be useful when it is impractical to decrease the solvent power but when a better separation is required. This is shown in Fig. 3–5 in a liquid–solid chromatographic separation with the samples run at 3° and 24°C. Separations of such compounds as steroids (Siggia and Dishman, 1970) can, however, be obtained at ambient temperatures. Brooker (1971) found that high flow rates and pressures create excess noise when using the Varian micro-UV detector that absorbs at 254 nm. If high temperatures

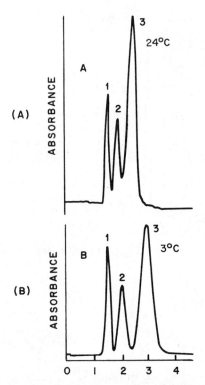

Fig. 3–5. Effect of temperature on liquid–solid chromatography. *Instrument*: Waters ALC 201. *Column*: 2 ft × 2.3 mm i.d.; packed with Carbowax 400/Porasil C. *Mesh*: 36–75. *Solvent*: isooctane. *Flow rate*: 1.5 ml/min. *Chart*: 0.75 in/min. *Samples*: (1) decahydronaphthalene; (2) *o*-quarterphenyl; (3) *m*-quarterphenyl; (4) 1,2-benzanthracene. [Contributed by Waters Associates, Inc. (1970).]

(70°–80°C) are also used, a still higher noise level in the detector occurs. He improved markedly the stability and sensitivity of the detector by modifying it so that the temperature of the effluent through the flow cell could be controlled. This modification allowed, with the use of higher flow rates (25–50 ml/hr), the assay of less than 15 pmoles of cAMP in biological samples.

In gel permeation chromatography a change in temperature does not significantly change the retention time. However, an increase in temperature may cause a decrease in retention time because the solvent becomes less viscous. Once temperature for a separation is optimized, it is important that it not be changed because variations in temperature may adversely affect accurate measurements. Since it is difficult to dissolve some polymers in a suitable solvent at room temperature, it may be necessary to maintain the injector, column, and detector at controlled, elevated temperatures.

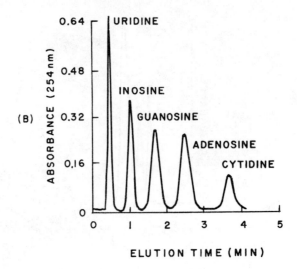

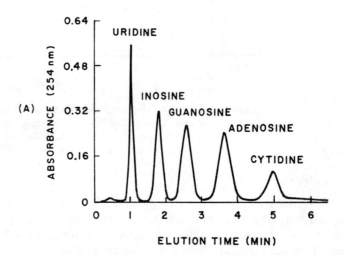

Fig. 3–6. Effect of particle size. (A) The rapid separation of ribonucleosides by high-pressure liquid chromatography with a 7- to 10μm conventional cation exchange resin. *Instrument*: Varian LCS 1000. *Column*: 0.24 × 25 cm. *Resin*: Bio-Rad Aminex A-7 (8×). *Eluent*: 0.4 mole/liter ammonium formate, pH 4.75. *Flow rate*: 35 ml/hr. *Pressure*: 4400 psi. *Temperature*: 55°C. *Sample*: 0.5 μg of each nucleoside.

(B) The rapid separation of ribonucleosides by high pressure liquid chromatography with a 3- to 7-μm conventional cation exchange resin. *Instrument*: Varian LCS 1000. *Column*: 0.24 × 25 cm. *Resin*: Durrum DC-× (8×). *Eluent*: 0.4 mole/liter ammonium formate, pH 4.75. *Flow rate*: 40 ml/hr. *Pressure*: 4800 psi. *Temperature*: 55°C. *Sample*: 0.5 μg of each nucleoside. (Contributed by Burtis *et al.* (1970c).)

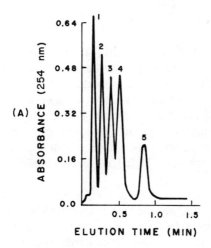

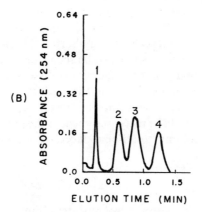

Fig. 3–7. Effect of column length. (A) The effect of column length on the separation of ribonucleosides by high pressure liquid chromatography. *Instrument*: Varian LCS 1000. *Column*: 0.24 × 15 cm. *Resin*: Durrum DC-x (8×), 3–7μ. *Eluent*: 0.4 mole/liter ammonium formate, pH 4.50. *Flow rate*: 60 ml/hr. *Pressure*: 4500 psi. *Temperature*: 85°C. *Samples*: 0.5 μg of each nucleoside: (1) uridine; (2) inosine; (3) guanosine; (4) adenosine; (5) cytidine. [Contributed by Burtis *et al.* (1970c).]

(B) The rapid separation of ribonucleosides by high pressure liquid chromatography with a 7- to 14-μm conventional cation exchange resin. *Instrument*: Varian LCS 1000. *Column*: 0.24 × 24 cm. *Resin*: Sondell VC-10 (10×, 7–14 μm). *Eluent*: 0.4 mole/liter ammonium formate, pH 4.50. *Flow rate*: 85 ml/hr. *Pressure*: 4900 psi. *Temperature*: 85°C. *Samples*: 0.5 μg of each nucleoside: (1) uridine; (2) guanosine; (3) adenosine; (4) cytidine. [Contributed by Burtis *et al.* (1970c).]

B. STATIONARY PHASE

As was discussed in Chapter 2, the stationary phase best suited for high pressure liquid chromatography is a column packing of solid microspheres with a porous surface. The type of sorbant depends on the separations desired. For ionic compounds, such as nucleotides, purines, pyrimidine bases, or amino acids, coatings of anion or cation exchangers are necessary. For separating nonpolar compounds by liquid–solid chromatography, solid packings which have porous, silica surfaces can be used; in liquid partition chromatography, packings which have been coated with a liquid can be utilized.

In deciding on the right column for the separation desired, some of the variables to be evaluated are the length and internal diameter of the column and the type of packing. In the packing, the size of the particle, the thickness of the shell (if a solid core support is used), the type of coating, and its polymeric structure must all be taken into consideration. Some of the supports available commercially are listed in Table 2–2. An example of the effect of particle size is shown in the separation of nucleosides and bases in Fig. 3–6. It should be noted, however, that since the flow rate was increased, the decrease in retention time is not completely due to particle size. In Fig. 3–7, faster separation was achieved using a shorter column. The effect of the shorter column is evident since it can be seen from the comparison of the two chromatograms that a much higher flow rate was required to obtain the separation in approximately the same time using a longer column. From these illustrations it can be seen that for faster separation there are two options: the flow rate could be increased because the particle size was smaller or the length of the column could be reduced.

C. TYPES OF ELUENTS

The type of liquid chromatography determines whether organic or aqueous eluents are to be used in liquid chromatography. This in turn is determined by the type of separation to be achieved and the kind of column utilized. Aqueous solutions are used in ion exchange chromatography and either aqueous or organic solvents or both are used in partition, adsorption, or gel permeation chromatography.

In using a liquid chromatograph with a UV detector it is important that neither the eluents nor any impurities in them absorb strongly in the UV at the wavelength used in the detector. The best type of eluents to be used in separating a group of compounds, must be determined for each specific case. However, it is possible to adapt information gained in the use of conventional column chromatography to high pressure systems. Usually both the concentrations of solutions and the volumes needed are greatly reduced. Formate solutions are widely used in classical column chromatography; however, in the high pressure systems they are not used if the separation can possibly be obtained with another anion because the formate may attack the stainless steel fittings. Therefore, if formate is used, the hardware must be examined regularly to prevent leaks in the system which cause pressure losses. Potassium dihydrogen phosphate has been the eluent of choice in use with highly sensitive UV detectors in liquid chromatographic systems because of its low inherent UV absorption. However, it has been found that reagent grade potassium dihydrogen phosphate contains enough UV-absorbing impurities to cause a rising baseline with increasing concentration of the gradient (Fig. 3–8). Shmukler (1970b) developed a method

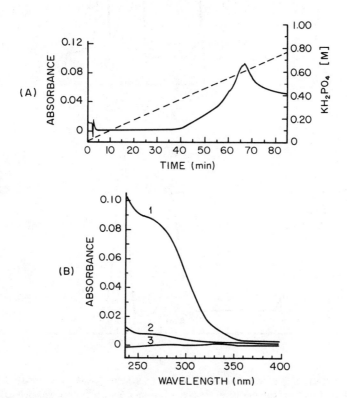

Fig. 3–8. UV absorbants in KH_2PO_4. (A) Chromatographic record of buffer only obtained with a linear gradient (dashed line) prepared from reagent grade KH_2PO_4. Chromatographic conditions: Dilute buffer, 15 ml 0.01 M; concentrated buffer, 1.20 M; flow in, 6 ml/hr; flow out, 12 ml/hr; photometer setting, 0.16 full scale deflection.

(B) UV spectra of 1.0 M KH_2PO_4 in water (1) reagent grade, (2) purified preparation, (c) water–water baseline. [Contributed by Shmukler (1970b).]

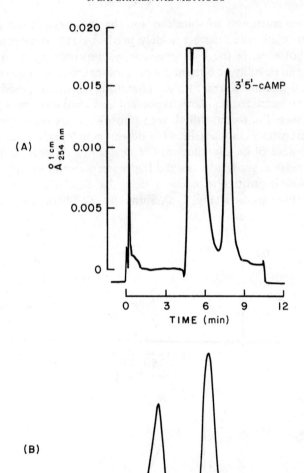

Fig. 3–9. (A) 3′,5′-cAMP. *Instrument*: Varian LCS 1000. *Sample*: 3′,5′-cAMP. *Inlet pressure*: 400 psi. *Flow rate*: column 10 ml/hr. *Column*: 1 mm × 300 cm; packed with pellicular anion exchange resin. *Temperature*: 80°C. *Eluent*: 0.01 M HCl, pH 2.1. *Attenuation*: 0.02 AU full scale. [Contributed by Brooker (1970).]

(B) *Instrument*: Varian LCS 1000. *Sample*: 5′-AMP and 3′,5′-cAMP. *Flow rate*: 12 ml/hr. *Column*: 1 mm × 300 cm; packed with pellicular anion exchange resin. *Eluents*: 0.015 M KH₂PO₄, pH 4.5; 0.25 M KH₂PO₄ in 2.2 M KCl, pH 4.5. *Attenuation*: 0.16 AU. *Starting volume*: 50 ml. [Contributed by Brown (1970).]

for purifying potassium dihydrogen phosphate; contaminants were removed by passage of the solution over a Dowex 1-8x phosphate column so that the phosphate solution can be used routinely in the separation of nucleotides and other compounds. Another gradient eluent system used by Shmukler (1970b) was one in which he increased the ionic strength of the high concentration phosphate solution by the addition of KCl instead of increasing merely the phosphate concentration. Better resolution was obtained without the positive drift in the baseline caused by impurities in the potassium dihydrogen phosphate. It was found, however, by both Brown (1970) and Shmukler (1970b) that a falling baseline may be obtained when this gradient was used which can cause problems at very low sensitivities. However, stable baselines have been obtained with this gradient. Examples of the rising, falling, and steady baselines are illustrated in Fig. 3–9. Other salts that have been used successfully with ion exchangers are the sodium, potassium, and ammonium salts of citrates, borates, acetates, nitrates, or sulfates. Dilute solutions of nitric acid, citric acid, phosphoric acid, and hydrochloric acid also have been used to obtain separation of specific compounds. Completely different eluent systems can be used for the chromatography of the same compounds. For example, Brown (1970) used a gradient of 0.015 M KH_2PO_4 and 0.25 M KH_2PO_4 in 2.2 M KCl to separate 3′,5′-cAMP from 5′-AMP and Brooker (1970) 0.10 M HCl for routine analysis of picomole amounts of 3′,5′-cAMP. This is shown in Fig. 3–10. For separating aromatic compounds and steroids by partition and adsorption chromatography, mixtures of water and organic solvents can be used.

D. TYPES OF ELUTION

1. Isocratic Elution Mode

In high pressure liquid chromatography either an isocratic or gradient elution mode may be used. The isocratic elution mode is desirable if it can possibly be used to achieve the separation because it makes it unnecessary to regenerate the column. In using the straight elution method, different separations may be obtained by varying the pH, ionic strength, or type of exchanging ion in the solution. For example, Brown (1970) found that AMP and GMP can be separated in 15 minutes using an eluent of 0.10 M KCl. Using an eluent of higher ionic strength (0.019 M KH_2PO_4 in 0.19 M KCl), ATP can be well separated from GTP in approximately the same amount of time. Therefore, it can be seen that the less polar compounds, the monophosphate nucleotides, require an eluent of lower ionic strength, whereas for the triphosphate mononucleotides, which are more highly charged, higher ionic strength is necessary. It also demonstrates the use of different

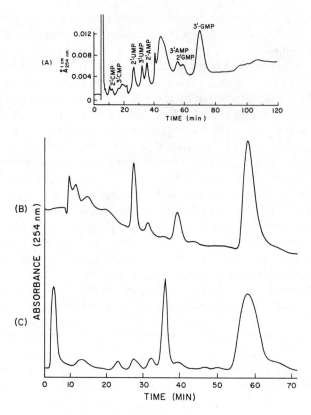

Fig. 3–10. (A) Rising baseline. *Instrument*: Varian LCS 1000. *Sample*: tRNA hydrolysate. *Flow rates*: 12 ml/hr; 6 ml/hr. *Column*: 1 mm × 300 cm; packed with pellicular anion exchange resin. *Temperature*: 80°C. *Eluents*: 0.01 M KH$_2$PO$_4$, pH 3.35; 1.0 M KH$_2$PO$_4$, pH 4.3. *Gradient delay*: 7.5 min. *Starting volume*: 50 ml. *UV output*: 0.16 AU. (Contributed by A. C. Griffin, Anderson Hospital for Tumor Research, Houston, Texas.)

(B) Falling baseline. *Instrument*: Varian LCS 1000. *Column*: 1 mm i.d. × 300 cm; packed with pellicular anion exchange resin. *Temperature*: 75°C. *Flow rate*: 12 ml/hr. *Initial volume*: 50 ml. *UV attenuation*: 0.04 AU. *Sample*: nucleotide extract of human erythrocytes plus 1 μl of UDPG soln. [Contributed by Brown (unpublished data).]

(C) Steady baseline. *Instrument*: Varian LCS 1000. *Column*: 1 mm i.d. × 300 cm; packed with pellicular anion exchange resin. *Temperature*: 75°C. *Flow rate*: 12 ml/hr. *Initial volume*: 50 ml. *UV attenuation*: 0.16 AU. *Sample*: nucleotide extract of pig whole blood. [Contributed by Brown *et al.* (1972).]

ions in the eluent to achieve good separation of differently charged species. If, however, the mono-, di-, and triphosphate nucleotides are to be separated in one chromatogram, because of the large differences in equilibrium constants of the three classes of compounds, a gradient elution method must

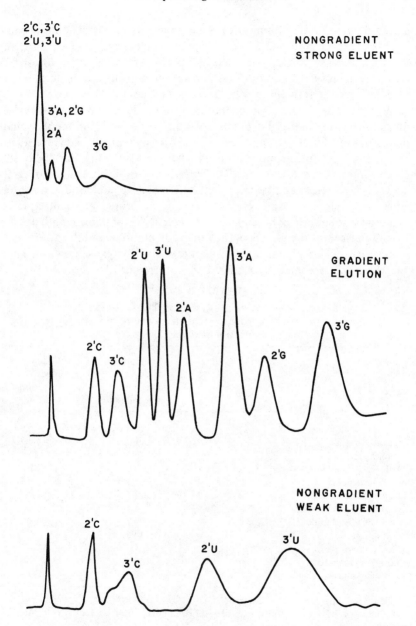

Fig. 3–11. Comparison of gradient and straight elution. *Instrument*: Varian LCS 1000. *Column*: 1 mm i.d. × 300 cm; packed with pellicular anion exchange resin. *Temperature*: 80°C. *Flow rates*: 12 ml/hr; 6 ml/hr. *Initial volume*: 40 ml. *UV attenuation*: 0.32 AU. [Contributed by Burtis and Gere (1970).]

be used. It was found by Shmukler (1970b) that, whereas the KCl alone could separate the monophosphate or diphosphate nucleotides or one group from the other, it was necessary to have the potassium dihydrogen phosphate present to achieve resolution of all three of these groups of nucleotides. At times, however, a gradient may be necessary to separate very closely related compounds. The comparison of the use of a gradient and straight elution modes can be seen in Fig. 3–11 in the separation of 2′- and 3′-monophosphate nucleotides of adenine, guanine, uracil, and cytosine (Burtis and Gere, 1970). With a straight elution mode using a strong eluent, only the 2′- and 3′-guanine nucleotides are separated. There is no resolution of the other 2′- and 3′-nucleotides. Using a weak eluent, only 4 of the 8 nucleotides are eluted in the same length of time. However, using the proper gradient elutions, good separation of all eight peaks is obtained in a reasonable length of time. The comparison of the effects of the nongradient mode and the gradient on retention times for glucuronide and sulfate conjugates is shown in Table 3–4, which was prepared by Anders and Latorre (1971).

TABLE 3–4

EFFECT OF GRADIENT ON RETENTION TIMES: RETENTION
VALUES OF PHENOLS AND THEIR CORRESPONDING
GLUCURONIDE AND SULFATE CONJUGATES[a]

Compound	Nongradient mode, retention time (min)[b]	Gradient mode, retention time (min)[c]
Phenol	5.0	5.0
Catechol	4.7	4.7
p-Nitrophenol	15.5	13.9
p-Hydroxyacetanilide	3.6	3.4
Phenyl glucuronide	3.1	9.7
Catechol glucuronide	2.7	3.1
p-Nitrophenyl glucuronide	4.5	3.7
Acetanilide glucuronide	2.7	8.5
Phenyl sulfate	12.5	—[d]
Catechol monosulfate	14.7	—[d]
p-Nitrophenyl sulfate	38.5	—[d]
Acetanilide sulfate	9.5	37.5

[a]Contributed by Anders and Latorre (1971).
[b]Conditions: 10.0 mM formic acid, pH = 3 containing 1.0 M potassium chloride, 80°C, 30 ml/hr.
[c]Conditions: low concentrate chamber, 1.0 mM formic acid, pH = 4; high concentrate chamber, 1.0mM formic acid, ph = 4 containing 2.0 M potassium chloride. Flow of high concentrate into gradient chamber: 15 ml/hr. Column flow: 30 ml/hr. Initial volume: 50 ml. Temperature: 80°C.
[d]No compound eluted within 90 minutes.

2. Gradient Elution

Gradient elution or solvent programming is the change of solvent composition during a separation in which the solvent strength increases from the beginning to the end of the separation. It is well suited to the analysis of samples of unknown complexity since good resolution is automatically provided for a wide range of sample polarities. Gradient elution has been used extensively in adsorption and ion exchange chromatography and is finding application in partition chromatography. Although it is analogous to temperature programming in gas chromatography, it is generally more powerful than temperature or flow programming for controlling capacity ratios in liquid chromatography. Snyder (1965) has reviewed comprehensively the subject of gradient elution.

There are two types of gradient systems: low-pressure gradient mixers and high-pressure gradient mixers. In the former the solvents are mixed at atmospheric pressure and then pumped to the column. In the latter, solvents are pumped into a mixing chamber at high pressure before going into the column.

For example, in the Varian LCS 1000, in order to form a gradient, the gradient chamber is first filled with a predetermined volume of dilute eluent from the low-concentration eluent reservoir. Using a pump, the concentrated buffer is pumped into the gradient chamber. This is continually stirred. The resultant gradient is then pumped through the column which is capable of leak-free operations at the pressure desired. A linear gradient is one in which the column flow rate is twice that of the flow rate of the high concentration eluent into the low. The slope of the linear gradient is determined by the initial volume of the dilute eluent in the gradient chamber and can be varied by changing this volume. By decreasing the volume of low concentration eluent and thus by increasing the slope of the linear concentration gradient, a faster analysis may be obtained. By varying the flow rates of the two pumps, a linear, concave, or convex gradient is obtained. In gradient elution the composition of the mobile phase changes as a function of time. Therefore, gradient elution is preferred in many separations to obtain better resolution, shorter analysis time, or to improve peak shape. In a gradient, as in an isocratic elution mode, different separations can be obtained by varying the pH, the ionic strength, or the type of exchanging ion of the low concentration or high concentration eluent or both. R. P. W. Scott (1971a) has developed a simple computer program to provide curves which relate the mobile phase composition with time for the incremental method of mixing.

It is also possible to delay the start of the gradient so that certain compounds in a sample can be separated using a nongradient elution mode. Other compounds in the sample are subsequently eluted with a gradient.

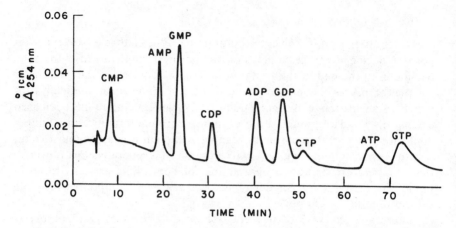

Fig. 3–12. Effect of eluent on retention time. *Instrument*: Varian LCS 1000. *Sample*: mono-, di-, and triphosphates of nucleosides. *Inlet pressure*: 400 psi. *Flow rate*: column 12 ml/hr; gradient 6 ml/hr. *Column*: 1 mm × 300 cm; packed with pellicular anion exchange resin. *Temperature*: 75°C. *Starting eluent*: 0.25 M KH$_2$PO$_4$ in 2.2 M KCl, pH 4.5. *Gradient delay*: none. *Initial volume*: 50 ml. *Attenuation*: 0.08 AU full scale. [Contributed by Brown (1970).]

In separating nucleotides, changing the pH of the eluent and the temperature affected the resolution, but to a much lesser degree than the slope of the linear concentration gradient or the ionic strength of the eluents (Brown, 1970). The effect of pH on retention time and resolution are shown in Table

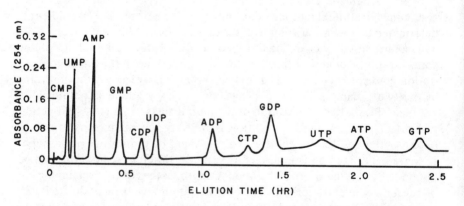

Fig. 3–13. Effect of eluent on retention time. *Instrument*: Varian LCS 1000. *Column*: 1 mm × 300 cm; packed with pellicular anion exchange resin. *Temperature*: 80°C. *Sample*: mono-, di-, and triphosphates of the nucleosides. *Inlet pressure*: 2000 psi. *Flow rate*: column 24 ml/hr. *Gradient*: 12 ml/hr. *Starting eluent*: 0.01 M KH$_2$PO$_4$, pH 3.25. *Gradient eluent*: 1.0 M KH$_2$PO$_4$, pH 4.2. *Gradient delay*: 7.5 min. *Initial volume*: 40 ml. *Attenuation*: 0.32 AU full scale. [Contributed by Burtis and Gere (1970).]

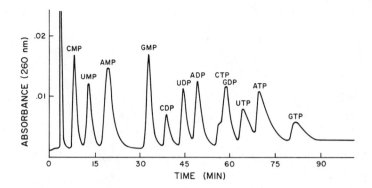

Fig. 3–14. Effect of eluent on retention time. *Instrument*: noncommercial. *Column*: 1.0 mm i.d. × 193 cm; packed with pellicular strong anion exchanger. *Eluent*: linear gradient of ammonium formate buffer, pH 4.3, from 0.04 M to 1.5 M. *Temperature*: 71°C. *Flow rate*: 12 ml/hr. *Pressure*: 51 atm. [Contributed by Horvath, *et al.* (1967).]

3–5. Examples of the effects of different ionic strengths and types of ions are shown in nucleotide chromatograms of Brown (1970) (Fig. 3–12), Burtis (1970) (Fig. 3–13), and Horvath (1972b) (Fig. 3–14). The same stationary phase was used. In separating the adenine, guanine, cytosine, and uracil mono-, di-, and triphosphate mononucleotides, Burtis used 0.01 M KH_2PO_4 and 1.0 M KH_2PO_4, and Brown used 0.015 M KH_2PO_4 and 0.25 M KH_2PO_4 in 2.2 M KCl. Horvath used a linear gradient of ammonium formate buffer from 0.04 M to 1.5 M, pH 4.35. This analysis was run at 71°C compared to 75°C used by Brown and 80°C by Burtis and Gere. It should be noted the reversal of position of the CTP and GDP peaks in Horvath's chromatogram as compared to the other two. Even with increasing the flow rate of the linear concentration gradient to 24 ml/hr, the analysis run by Burtis required $2\frac{1}{2}$

TABLE 3–5

EFFECT OF pH ON RETENTION TIME AND RESOLUTION[a]

	Retention time (min)				Resolution	
	2'-CMP	3'-CMP	2'-AMP	3'-AMP	2' + 3'-CMP	2' + 3'-AMP
Normal gradient	13	18	28	36	2.5	2.0
Gradient with pH 3.35	31	31	41	47	0.0	1.5
Gradient with pH 4.3	10	14	23	33	2.1	2.0

[a]Contributed by Burtis and Gere (1970).

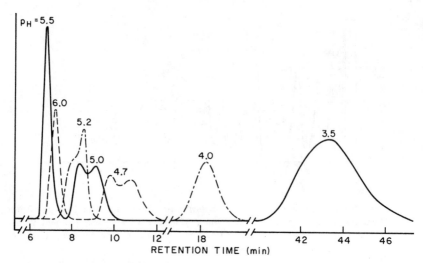

Fig. 3–15. Effect of pH on elution pattern of guanines. *Instrument*: Varian LCS 1000. *Column*: 1 mm i.d. × 300.7 cm; packed wih peculĺiar cation exchange resin. *Eluent*: 0.02 *M* KH₂PO₄. *Flow rate*: 14.8 ml/hr. *Temperature*: 60°C. [Contributed by Horvath (1972a).]

hours, whereas that of Brown using a flow rate of 12 ml/hr, needed only 70 minutes. Horvath, using a different eluent system was able to obtain the separation in about 80 minutes. It is interesting to note that this separation by conventional ion exchange chromatography when first reported required 30 hours for completion.

Determination of the optimum pH range is important with many compounds. In separating purine and pyrimidine bases, Horvath and Lipsky (1969b) found that increasing the pH reduced the retention times. They also found that the spurious peaks observed at low pH were related to the ionization of the solutes. An example of the effect of pH of the eluent on the retention time and shape of guanine peaks is shown in Fig. 3–15.

Uziel and Koh (1971) found that in order to elute minor nucleosides and bases of RNA hydrolysates, it was necessary to incorporate ethanol into aqueous solvents. These nucleosides, such as 2-methyl-thio-N⁶-isopentyladenosine, N⁶-isopentyladenosine, and 7-methylguanosine, which are strongly positively charged and highly hydrophobic, were completely recovered using a cation exchange column and an eluent which had a final composition of 0.85 *M* ammonium acetate (pH 5.7) and 15% ethanol.

In liquid–solid chromatography and partition chromatography as well as in ion exchange chromatography, solvent programming is very useful (Fig. 3–16). The eluents are not limited to salt solutions, as in the case of ion

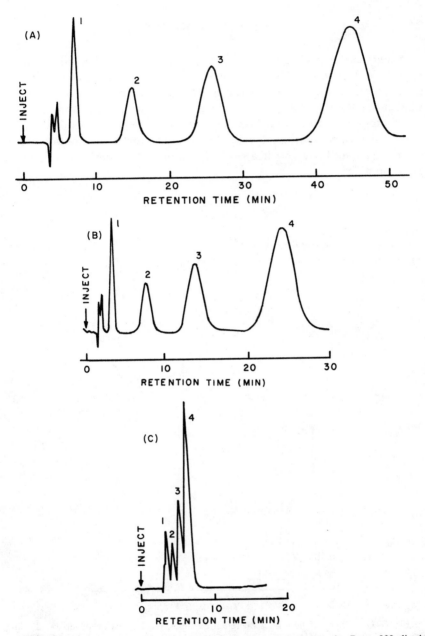

Fig. 3–16. Effect of eluent on retention time. *Instrument*: du Pont 820 liquid chromatograph. *Column*: 1 m × 2.1 mm i.d. *Support*: hydrocarbon polymer. *Mobile phase*: (A) 75% H_2O/25% methanol (v/v); (B) 50% H_2O/50% methanol (v/v); (C) 25% H_2O/75% methanol (v/v). *Pressure*: 1200 psig. *Detector*: UV photometer at 254 nm. *Samples*: (1) naphthalene; (2) anthracene; (3) pyrene; (4) benz(α) anthracene. [Contributed by du Pont Instruments (1969).]

exchange chromatography, and organic solvents may be programmed. Snyder and Saunders (1969) reported that a logarithmic concentration gradient with respect to time, gave the best separations. R. P. W. Scott and Lawrence (1970) described the advantage of polarity programming in liquid chromatography. In this method, a gradient elution is obtained under conditions of axial equilibrium. It relies on the change of the distribution coefficient of alcohol between silica gel and a nonpolar solvent by temperature programming.

E. FLOW RATES

In considering the flow rate of a nongradient elution mode, the flow rate of the eluent into the column can be varied to obtain the best elution in the

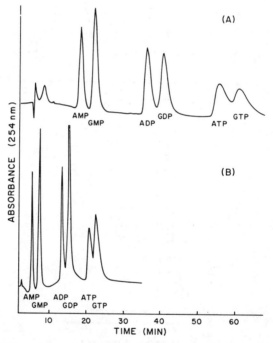

Fig. 3–17. Effect of flow rates on retention times. (A) Separation of mono-, di-, and triphosphate nucleotides of adenosine and guanosine. *Instrument*: Varian LCS 1000. *Starting volume*: 50 ml. *Flow rates*: 12 and 6 ml/hr. *Eluents*: 0.015 M KH$_2$PO$_4$ and 0.25 M KH$_2$PO$_4$ in 2.2 M KCl, *Samples*: 3μl of a mixture of 0.5 mM AMP, ADP, and ATP and 3μl of a mixture of 0.5 mM GMP, GDP, and GTP. *UV output*: 0.08 AU.

   (B) *Instrument*: Varian LCS 1000. *Starting volume*: 50 ml. *Flow rates*: 24 ml/hr; 12 ml/hr. *Eluents*: 0.015 M KH$_2$PO$_4$ and 0.25 M KH$_2$PO$_4$ in 2.2 M KCl. *Samples*: 3 μl of a mixture of 0.5 mM AMP, ADP, and ATP and 3 μl of a mixture of 0.5 mM GMP, GDP, and GTP. [Contributed by Brown (1970).]

shortest possible time. This is particularly true in routine analyses, such as used by pharmaceutical, food, or flavor companies who are monitoring the impurities that may be present in a product. It is also utilized in research laboratories where the amount of one product or a group of similar compounds in a sample will be determined over and over again. For example, the straight elution mode was used by Brooker (1970) to determine routinely the amount of 3',5'-cAMP in cell extracts.

In the determination of the optimum flow rate, several factors must be

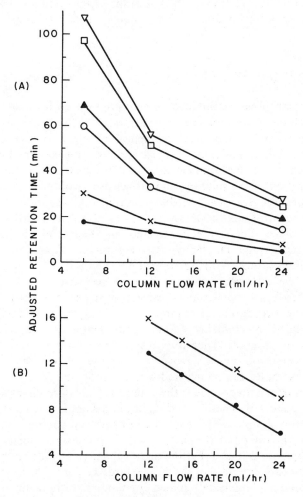

Fig. 3–18. Graph of retention time vs flow rates. *Instrument*: Varian LCS 1000. *Column*: 1 mm × 3 m; packed with pellicular anion exchanger. *Eluents*: 0.015 M KH$_2$PO$_4$, 0.025 M KH$_2$PO$_4$ in 2.2 M KCl. *Sample*: AMP and GMP. ($\triangledown$) GTP; ($\square$) ATP; ($\blacktriangle$) GDP; ($\bigcirc$) ADP; ($\times$) GMP; ($\bullet$) AMP. [Contributed by Brown (1971a).]

considered: first, the resolution desired; second, the rapidity required of the analysis; and third, the number of compounds to be separated. For example, when only six solutes, such as the adenine and guanine nucleotides, are present in a solution, the time of analysis can be cut in half and better resolution obtained if the flow rate of the linear gradient is doubled (Fig. 3–17). However, this flow rate is not practical if the cytosine, uracil, and thymine nucleotides are also present and their concentrations are to be measured since there is not space in the chromatogram for good resolution of 15 peaks (Fig. 3–12) (Brown, 1970). A graph of the effect of flow rates on adjusted retention time of adenine and guanine nucleotides is shown in Fig. 3–18.

F. SAMPLE SIZE AND INJECTION

In most high pressure liquid chromatographs used for analytical purposes, the volume of solution and concentration of sample are very low and generally much smaller than that of comparable conventional column chromatography. The volume used in most systems ranges from 1 to 30 μl, depending on the design and sensitivity of the instrument, the column, the concentration of the solution, and the sensitivity of the detector for the solutes in question. Samples must be measured very accurately since with such low concentrations the loss or addition of an extra microdrop can cause inaccuracies. The sample is injected into the instrument with a carefully calibrated syringe. The sample introduction technique of the operator is of utmost importance. With the stopped-flow method, the sample injection technique is easier because there is no back pressure to cause a loss of sample. Moreover, if the system is properly designed, the needle fits right down into the precolumn or column itself. To prevent sample loss, the sample should be injected slowly since all the sample may not be absorbed rapidly into the column. Therefore, a pool may form at the inlet if the sample is injected too quickly. However, with the new injection loop techniques, large samples can be injected with no difficulty.

If the column is overloaded, the retention times are changed and the peak shapes distorted. Therefore, the optimum sample size must be determined for each sample under each set of operating conditions. R. P. W. Scott (1971b) investigated the factors that affect peak capacity in liquid chromatography and found that the peak capacity depends on the column efficiency and capacity ratio of the most retarded solute. The loading capacity of the column is largely determined by column size (diameter and length) and the type of packing material. For example, the sample capacity of pellicular ion exchange columns is generally smaller than that of comparable columns containing conventional ion exchange resins. This is due to the fact that there is a relatively small amount of resin present in the column. The volume is occupied largely by the inert support, the microglass beads.

G. DETECTION

It is important that the correct detector be chosen for the samples to be analyzed. Whereas the UV detector is well suited for work with nucleotides because they have large molar absorptivity (10^4) at 254 nm, it is not suitable for all work. It is not possible to detect compounds which have no absorbance at this wavelength unless derivatives can be made that do absorb. If it is necessary to analyze for every solute in the sample, the refractive index detector or microadsorption detector may be preferable. They are not, however, as sensitive as the micro-UV detector now on the market and allowances for the difference in sensitivity must be made. However, it is possible to have an instrument supplied with two or more types of detectors. Any one or a combination could be used for the analysis. It is important that each operator know not only the possible applications of the detector he is using but also its limitations as well. The use of both a refractive index and a UV detector are shown in Fig. 3–19. It is also possible to use two UV detectors at different wavelengths. An example of this is shown in Fig. 3–20.

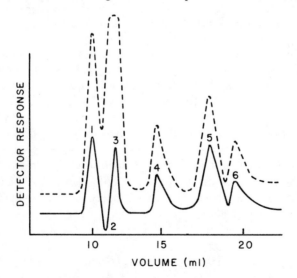

Fig. 3–19. Use of dual UV and RI detectors. *Instrument*: Waters ALC-100. *Column*: 12 ft × 2.3 mm. *Support*: Durapak-Carbowax 400/Porasil C, 37–75 μm. *Temperature*: 23°C. *Solvent*: chloroform. *Samples*: (1) caffeine; (2) unknown; (3) phenacetin; (4) aspirin; (5) salicylic acid; (6) salicylamide. (—) Differential refractometer; (---) UV at 254 nm. [Contributed by Waters Associates, (1970).]

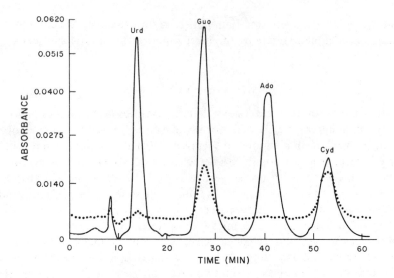

Fig. 3–20. Use of dual wavelengths in UV detector. *Instrument*: Varian LCS 1000. *Column*: 6 mm × 19 cm; Bio-Rad Aminex A-6 cation exchange resin. *Eluent*: 0.4 ammonium formate. *Flow rate*: 24 ml/hr. *Temperature*: 50°C. *Samples*: (1) uridine; (2) guanosine; (3) adenosine; (4) cytidine. (—) 260 nm; (. . .) 290 nm. [Contributed by Horvath (1972a).]

H. CALCULATION AND RESULTS

Since there are many ways to quantitate the results, the best possible way available should be chosen. The quantitative results should be similar in precision and accuracy to those found in gas chromatography. It is preferable that the instrument be calibrated for each particular analysis with known quantities of the sample components.

Because of the effect on peak areas (Brown, 1971a), flow rates must be kept constant (Fig. 3–18). Also, for good accuracy, temperatures must be maintained. Since it is known that temperature, flow rate, slope of the linear concentration gradient, ionic strength, and pH of the eluents all affect retention times, these operating conditions must be kept consistent for the best possible results.

III. Preparation of Sample

Since sample preparation also influences the quantitative results, the best possible method must be worked out for each individual sample and great care taken in preparing it. An example of the differences in nucleotide

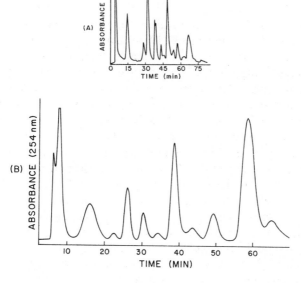

Fig. 3–21. (A) Effect of extraction procedures. *Instrument*: noncommercial. *Column*: 1 mm i.d. × 193 cm; packed with pellicular anion exchange resin. *Flow rate*: 12 ml/hr; 6 ml/hr. *Eluent*: linear gradient of ammonium formate 0.04 *M* to 1.5 *M*, pH 4.35. *Temperature*: 71°C, *Sample*: cellular extract of mouse liver. [Contributed by Horvath *et al.* (1967).]

(B) Extraction procedures. *Instrument*: Varian LCS 1000. *Column*: 1 mm × 3 m; packed with pellicular anion exchange resin. *Flow rate*: 12 ml/hr.; 6 ml/hr. *Eluents*: 0.015 *M* KH_2PO_4; 0.25 *M* KH_2PO_4 in 2.2 *M* KCl. *Temperature*: 75°C. *Sample*: cellular extract of mouse liver. [Contributed by Clifford (unpublished data).]

profiles most possibly due to preparation of sample is the extract of mouse liver obtained several years ago (Horvath *et al.*, 1967) and that obtained recently by Dr. Andrew Clifford (Fig. 3–21 A and B). Although it is possible to analyze physiological fluids directly, it has been found that pretreatment of these fluids to remove proteins or other large molecules, prolongs the life of the column. C. D. Scott (1968b) and C. D. Scott *et al.* (1972) analyzed whole urine samples from normal male subjects; they found, however, that ultrafiltration of the urine, prior to chromatographic analysis to remove the protein, prolongs the life of the column. Burtis (1970b) used Hornig's procedure (1968) of prefractionation of the urine samples before analysis. For purine and pyrimidine bases, nucleoside or nucleotide analyses, samples may be prepared by removing substances which can interfere with the analysis or which can accumulate on the column. Compounds such as nucleic acids, proteins, and other large molecules can be removed by precipitation of these large molecules with cold acid, usually trichloracetic acid or perchloric acid, although cold $1N$ sulfuric acid has been used. The methods for extracting free nucleotides from tissues were summarized by Mandel (1964). Munro and Fleck (1964) reviewed extensively procedures for obtaining these components from nucleic acids and also for determining the free bases, nucleosides, and nucleotides in tissues. It must be emphasized that in order to obtain valid results of the amounts of free nucleotides in tissues, the method and technique of the preparation of the tissue is of utmost importance. This is especially true when using high pressure liquid chromatography because of its capacity for great sensitivity, accuracy, and precision. Since rapid degradation of the free nucleotides in tissue occurs, the preparation of the sample must be rapid to prevent degradation between the time when the animal is killed and that when the enzymes are inactivated. Therefore, the preparation of the tissue extract is critical if the subsequent identification and assay of the various kinds of free nucleotides is to give accurate results which correspond to the actual amounts of free nucleotides *in vivo*. Shmukler (1972a) found that there were great differences in the nucleotide profiles of rat brain depending on the method of preparing the extract. A high ATP/AMP ratio (100:1) was obtained if the rats were sacrificed by immersing the head first in liquid nitrogen. If, however, the animals were decapitated before immersion in the liquid nitrogen, the ATP/AMP ratio dropped to less than 1. This shows very rapid decomposition of the ATP. Another important precaution, if free nucleotides are to be determined, is that the DNA or RNA is removed completely and that none is degraded in the extraction process. On the other hand, if the hydrolysates of DNA or RNA are to be measured, it is equally important that all the free nucleotides present be removed first so that the subsequent assay will be accurate. Miech and Brown (1972) investigated several techniques for the extraction of nucleotides and found that, for studies involving many samples, the quickest method which was quantitatively accurate was that of extracting with TCA and neutralizing with Tris.

IV. Use of Fraction Collector

A fraction collector can be readily put on-line with a high pressure liquid chromatograph. Most instruments are set up with a collection port to make this hookup easy and convenient. Fractions can then be collected automatically by time or drops. Therefore, the use of radioisotopes is facilitated. In many investigations, such as the study of the metabolic path-

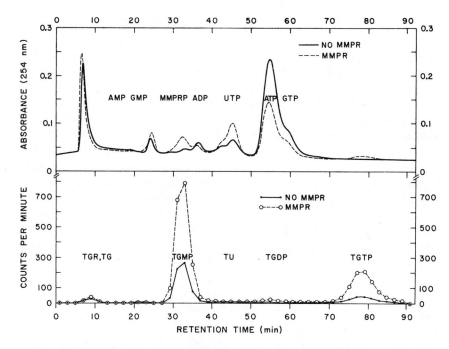

Fig. 3–22. Use of fraction collector. *Instrument*: Varian LCS 1000. *Column*: 1 mm × 3 m; packed with pellicular anion exchange resin. *Flow rates*: 12 ml/hr; 6ml/hr. *Eluents*: 0.015 M KH_2PO_4; 0.25 M KH_2PO_4 in 2.2 M KCl. *Starting volume*: 50 ml. [Contributed by Nelson and Parks (1972).]

ways of antimetabolites in cancer, the phosphorylation of the antimetabolites can be followed by two methods. This is illustrated in the work of Nelson and Parks (1972) who monitored the metabolism of thioguanine and 6-methyl-mercaptopurine riboside in Sarcoma 180 cells (Fig. 3–22).

V. Automation

Although the commercial liquid chromatographs now available are not automated (except for the amino acid analyzers), automation is becoming more and more necessary so that sophisticated measurements can become part of routine analyses in both the research and clinical laboratories. Automation is possible in the injection procedures, in carrying out chemical reactions involved in separating or identifying compounds, in regenerating the column after a gradient or solvent programming, and in obtaining and analyzing the data. The use of integrators and computers for data processing is discussed in Chapter 5. Automatic sampling for gel permeation chromatography has been available for some time. However, sampling systems for other types of high pressure liquid chromatography are just becoming available.

In the determination of the structure and sequences of bases in RNA and DNA, Uziel and Koh (1971) have made an important contribution by constructing an instrument that can automatically and selectively remove the terminal nucleotide residue from a ribonucleic acid. The chemical and separation steps which occur in the instrument are quantitative and since the cycles can be repeated, sequential degradation is achieved. The instrument is capable of controlling the reaction time, temperature, and solution contents in three separate vessels. It dispenses fluids, transfers solutions between the four reactors, filters the contents of each reaction vessel, separates the products by dialysis, automatically analyzes the released base, collects the remaining sample, and concentrates the retentate by ultrafiltrations.

VI. Column Maintenance

In high pressure liquid chromatography, most packing materials are fairly stable and chemically inert. Therefore, columns do not usually deteriorate easily or rapidly. There are, however, some fairly unstable ion exchangers on the market and therefore columns should be checked under the desired operating conditions for stability. Moreover, no material is indestructible and columns must be treated with respect. An example of a chromatogram from a column that had been used continually for $2\frac{1}{2}$ years (12–18 hours a day, 6 days a week) and had started to lose its resolving power and the same sample from a new column run just a few days later is shown in Fig. 3–23. When a gradient is used, the columns must be regenerated or returned to its initial condition of activity by removing all the more polar solvent from the active sites. This is accomplished by flushing the system with the starting solvent for a specified length of time. It has been found that if the regeneration procedure is varied greatly, it will have an effect on retention times. It is desirable that high pressure liquid chromatography systems be designed so that optimum regeneration is accomplished in minimum time.

When it is found that reproducible chromatograms of standard solutions are not being obtained, the column should be washed for a period of time. The type of solution used for washing must be determined for each set of conditions, depending on the type of separation, type of packing, and type of samples that have been analyzed. For example, with a pellicular anion ex-

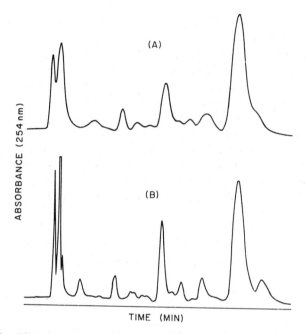

Fig. 3–23. Effects of resolving power of column on chromatograms. Both A and B run under the following conditions except that chromatogram A was from a column that had been used for $2\frac{1}{2}$ years and had begun to lose its resolving power (especially ATP from GTP) and B was from a new column. *Instrument*: Varian LCS 1000. *Column*: 1 mm × 3 m; packed with pellicular anion exchange resin. *Flow rates*: 12 ml/hr; 6 ml/hr. *Eluents*: 0.015 M KH$_2$PO$_4$; 0.25 M KH$_2$PO$_4$ in 2.2 M KCl. [Contributed by A. J. Clifford and P. R. Brown (unpublished data).]

changer, it has been found that flushing overnight with 0.1 N HCl, followed by water and the low concentration eluent, was most effective in bringing back the nucleotide separation power of a column that does not give the resolution it originally had. In other columns, basic solutions may be necessary to remove contaminants that have accumulated on the column. If the impurities have amassed over a long period of time, it may take several days for the cleaning process. It may also take time for the column to equilibrate and return to its original state. It is possible to clean a column outside of the instrument so that the latter can be used for research as much as possible.

In some instruments that are equipped with a precolumn, it is possible to replace the precolumn. The object of having a precolumn is to protect the column from contaminants.

In order to keep a column in good working order, it is important that it does not dry out so that there will be no change in the stationary phase and in the packing structure. Especially if a pellicular ion exchanger is used, the coating may crack, flake off, or deteriorate if allowed to overheat. Therefore, the column should not stay unused in a heated oven for any length of time. As in conventional column chromatography, since it is preferable that the column not be allowed to go dry, purging the column every day will keep it ready for use when needed, even if no samples are being run. If salt solutions are used, it is important to wash the column thoroughly with distilled water if the column is not going to be used for a period of time to prevent crystallization in the column and lines. Furthermore, since phosphates provide an excellent medium for mold growth, this type of contamination could cause severe problems in chromatographic separations. Leitch (1971), in an article on high speed liquid–liquid chromatography, advised that to promote long column life there should be true equilibration of the mobile liquid, prevention of stationary base oxidation, and proper temperature control.

VII. Instrument Maintenance

A. ROUTINE MAINTENANCE

Routine instrument maintenance and care will ensure that the chromatograph is operating correctly and will minimize "down time" or the time that the instrument is out of service. In order to get the best results and continuous trouble-free operation from a chromatograph, it is essential that the instrument be kept leak-free and clean. In most cases where salt solutions are used, leaks can easily be seen because of salt encrustation. If other solvents are used, the appearance of liquid denotes a problem. If the pressures and flows are lower than usual or the retention times longer, it is an indication of a leak in the system.

It cannot be overemphasized the importance of keeping the system clean. The eluents used should be free of suspended solids or particles since they can lodge in pumps and damage them, tubing, or the detector itself. Phosphate solution is an excellent medium for mold growth which can cause plugged systems and columns or unusual chromatograms. Solvent spills, especially with salt solutions should be cleaned up immediately to prevent corrosion.

The pumps should be checked and oiled regularly to ensure maximum operation.

B. LOG BOOK

A great help in instrument maintenance is the setting up and keeping of a log book for each instrument. Every analysis should be entered in the book

MACHINE B

NAME	DATE
SPECTRUM #	REFERENCE
SAMPLE	
VOLUME OF SAMPLE	STARTING VOLUME
MOLARITY OF SAMPLE	ELUENT
UV OUTPUT	FLOW RATE

Fig. 3–24. Stamp for recording necessary information on chromatogram.

with all information about the conditions used. The information should also be written on each chromatogram. For this purpose, a stamp can be made up so that even a novice operator will record the vital statistics necessary to provide at a glance a complete picture of the chromatogram. An example of such a stamp is shown in Fig. 3–24. For example, in our laboratory the spectrum number, date, sample, volume and molarity of sample, eluents, flow rates, UV output, and operator are stamped on each chromatogram. In the log book, pump gauge settings and UV zero are also included. Operating temperatures and pressures may also be noted. There is a space for comments so that we can tell at a glance whether a chromatogram was good or bad, when trouble spots started, how long they lasted and what the problem was. Entries also include the date of precolumn, column or UV lamp changes or any repairs that were made.

C. TROUBLE SYMPTOMS

When a liquid chromatograph is not operating correctly, it is easier to correct the problem if the symptoms are recognized promptly. For example, if there is no column flow or pressure, there may be a leak in the system before the column. If the column pressure is erratic, there may be a leak in the high pressure system before the column or there may be particles or air in the column pump check valves. If, however, the column pressure increases but the flow decreases, there may be a restriction in the precolumn or detector cell. Column overloading or adsorption of the sample on an ion exchange resin will cause poor peak shape. A loss of resolution is an important symptom and can have several causes: overloading, deteriorated column packing, strong retention of a solute on an ion exchange column, loss of the liquid phase from a column, or a leak in the system. An equally important symptom is the increase of retention time for known solutes. This may be caused by a decrease in solvent flow rate, change in temperature, poor gradient control, increasing activity on adsorption columns, or loss of the liquid phase from a column. If, on the other hand, the retention times are decreased, either the column was not regenerated after the gradient elution, or again there was a loss of the liquid phase from the column. The baseline should be monitored for possible problems and if there is no UV illumination, there may be a defective UV lamp, fuse, lamp starter, or loose electrical connection. When the recorder pen will not zero, there may be contamination in the cell, column bleed, bubble in the cell, or a gasket blocking the light paths. A noisy baseline can also be caused by contamination or air bubble in the cell, a defective UV lamp or recorder instrument grounding problem. Baseline drift may be a symptom of contamination or air bubble in the cell, contamination in the eluents, a contaminated or "bleeding" column or a leak between the sample and reference cells. Spikes on the recorded baseline may be due to bubbles passing through the cell. If there are negative peaks on the recorder, the photodetector assembly or flow cell assembly may be reversed.

It is advisable to keep a stock of spare parts so that "down time" will be minimized. The instrument companies usually can supply a list of fittings and parts that are likely to be needed for repairs. Since the lifetime of a column or UV lamp cannot be guaranteed and since much depends upon their use, it is expedient to have an extra readily available.

In many cases, an alert operator or technician can mean the difference between good and poor performance of a liquid chromatograph. Although many people may run samples on an instrument, it is essential to have one person responsible for it. A person who is attuned to its performance can often spot potential problems before major trouble results.

CHAPTER 4

IDENTIFICATION OF PEAKS

Although chromatography is an excellent separation method, it is also a valuable analytical tool because the identification, quantitation, and collection of the peaks are possible. One of the most important goals in analyzing cell extracts or other solutions of unknown composition by high pressure liquid chromatography is the identification of peaks. Chromatographic identification is most often made by retention time or volume. The retention data are characteristic of the sample and the mobile phase if all other operating conditions remain the same. Eluent peaks can be identified in the following ways:

1. Comparison of retention data of unknowns to that of standard solutions
2. Use of an internal standard
3. Isotopic labeling
4. Collection of fractions and characterization of the eluent peaks by spectrophotometric, chemical, or other chromatographic methods
5. The enzymic peak-shift technique
6. Derivatization

I. Use of Retention Data

The uncorrected retention time depends on many factors: column dimension and length, stationary phase, column temperature, flow rate,

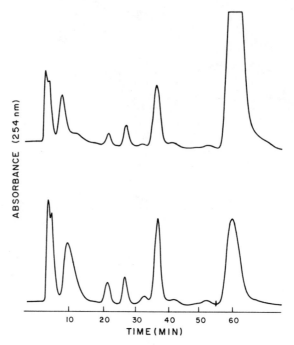

Fig. 4–1. Reproducibility of retention times. *Instrument*: Varian LCS 1000. *Column*: 1 mm × 3 m; packed with pellicular anion exchange resin. *Eluents*: 0.015 M KH$_2$PO$_4$, pH 4.5; 0.25 M KH$_2$PO$_4$ in 2.2 M KCl, pH 4.5. *Flow rates*: 12 ml/hr; 6 ml/hr. *Temperature*: 75°C. *Sample*: Nucleotide extracts of human erythrocytes. (Unpublished data by permission of Scholar *et al.*, 1972a.)

mobile phase, and instrument dead volume. Many compounds have not only characteristic retention times but shapes as well.

In a high pressure liquid chromatographic instrument, where operating conditions are maintained constant, there is good precision of retention data. An example of the reproducibility of data is illustrated in Fig. 4–1. The cell extracts of human erythrocytes, taken from two different male donors, were processed by two different researchers in two different laboratories, one using a TCA extraction method and the other a PCA technique. Both chromatograms were run on the same instrument under the same conditions. The only difference in the chromatograms was that in the upper one (Scholar) there was no UV attenuation for the ATP peak, whereas for the lower one (Clifford) the UV was attenuated so that the peak would not go off-scale. Therefore, it is possible to run standard solutions of the compounds which may be in a cell extract and compare the peaks in a chromatogram of the extract to those of known peaks analyzed under identical conditions. If the operating conditions are changed at all, new standard chromatograms

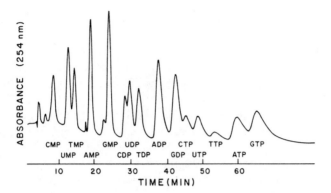

Fig. 4–2. Mono-, di-, and triphosphate mononucleotides. *Instrument*: Varian LCS 1000. *Column*: 1 mm × 3 m; packed with pellicular anion exchange resin. *Eluent*: 0.015 M KH$_2$PO$_4$, pH 4.5; 0.25 M KH$_2$PO$_4$ in 2.2 M KCl, pH 4.5. *Flow rates*: 12 ml/hr; 6 ml/hr. *Temperature*: 75°C. *Starting volume*: 50 ml. *Sample*: mixture of mono-, di-, and triphosphate mononucleotides. (Contributed by Brown (1970).)

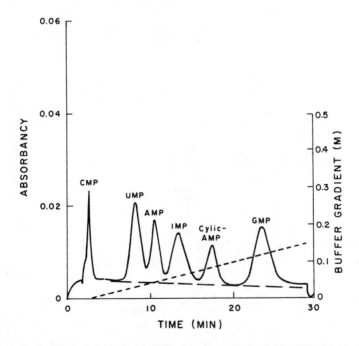

Fig. 4–3. Separation of AMP and GMP. *Instrument*: Varian LCS 1000. *Column*: 1 mm × 3 m; packed with pellicular anion exchange resin. *Eluents*: linear gradient of KH$_2$PO$_4$ from 0.01 to 0.16 M; starting buffer contained 0.01 M H$_3$PO$_4$. *Flow rates*: 24 ml/hr; 800–1000 psi. *Temperature*: 70°C. *Sample*: 1.01 nmoles CMP; 0.82 nmoles UMP; 0.86 nmoles AMP; 0.73 nmoles 3′,5′-cAMP; 0.56 nmoles IMP; 1.07 nmoles GMP. [Courtesy of Shmukler (1972).]

must be run. A standard chromatogram of the mono-, di-, and triphosphate ribonucleotides of cytosine, uracil, thymine, adenine, and guanine bases is shown in Fig. 4–2 (Brown, 1970). In analyzing for nucleotides, several compounds may have the same retention time, and other methods must also be used to identify the compound positively. This is especially true in the mono-phosphate nucleotide region, where GMP, FMN, UDPG, and CAMP have approximately the same retention times using the operating conditions of Brown (1970). Sometimes, it is possible to change the operating parameters to achieve better separation of compounds which have similar retention times; thus, optimizing the conditions to obtain the best possible resolution. Shmukler (1972b) modified the eluent system and was able to separate 3′, 5′-cAMP from GMP. In all previous analyses, the 3′, 5′-cAMP had the same retention time as GMP. Using this method, he was also able to obtain good separation of AMP from IMP (Fig. 4–3). Using a different eluent system, he

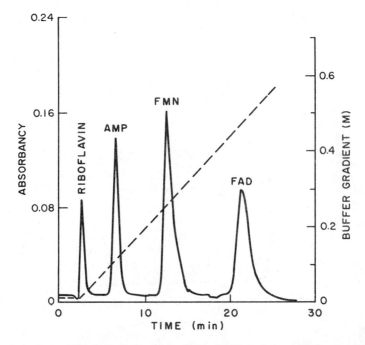

Fig. 4–4. Separation of riboflavin, AMP, FMN, and FAD. *Instrument*: Varian LCS 1000. *Column*: 1 mm × 3 m; packed with pellicular anion exchange resin. *Eluents*: 0.01 M KH$_2$PO$_4$ + 0.001 M H$_3$PO$_4$; 0.25 M KH$_2$PO$_4$ + 2.20 M KCl. *Flow rates*: F_1, 12 ml/hr; F_2, 24 ml/hr. *Temperature*: 70°C. *Sample*: 4.4 nmoles AMP; 5.81 nmoles FMN; 3.54 nmoles FAD. [Courtesy of Shmukler (1972b).]

TABLE 4–1

RETENTION TIMES (min) OF STANDARD NUCLEOTIDES SOLUTIONS[a,b]

Monophosphates		Disphosphates		Triphosphates	
CMP	10	NAD	8½	NADP	28
UMP	12	UDPG	25	CTP	46
TMP	14	UDP-man	25	UTP	50
IMP	18	UDP-gal	25	TTP	54
AMP	19	CDP	29	ITP	58
GMP	23	UDP	31	ATP	60
cAMP	23	TDP	33	GTP	65
FMN	23	IDP	37		
cGMP	30	ADP	38		
XMP	36	NADH	39		
		GDP	43		
		UDPGA	44		
		FAD	52		

[a]Brown (1970).

[b]*Instrument*: Varian LCS 1000 liquid chromatograph. *Starting volume*: 50 ml. *Flow rates*: 12 and 6 ml/hr. *Eluents*: 0.015 M KH$_2$PO$_4$ and 0.25 M KH$_2$PO$_4$ in 2.2 M KCl. *UV output*: 0.08. *Column*: 1 mm × 3 m; packed with pellicular anion exchange resin.

separated riboflavin, FMN, FAD, and AMP in less than 30 minutes (Fig. 4–4). A set of retention times of compounds which may play a role in the metabolism of purines and pyrimidines in cells are listed in Table 4–1. Some analogs of purine nucleotides have also been characterized by their retention times and are shown in Table 4–2. Some of these nucleotides are of interest because their bases or ribosides are used experimentally in cancer chemotherapy. Retention times of nucleosides and bases which were separated by Uziel *et al.* (1968), using cation exchange chromatography, are shown in Table 4–3. It is stressed that these retention times are applicable only to the operating conditions under which the samples have been analyzed. Differences in column packing, column length or diameter, condition of the column, flow rate, eluent, temperature, and slope of the linear concentration gradient all affect the separation and the retention data.

To identify the more than 150 peaks obtained by high pressure anion exchange liquid chromatography of human urine (C. D. Scott *et al.*, 1967; C. D. Scott, 1968), and Katz and Burtis (1969) determined the elution peaks of more than 100 compounds by the chromatography of reference compounds. These compounds were chosen because they are possible constituents of urine, they absorb in the UV, and they are commercially available. Some of the compounds thus identified and their elution volumes (in milliliters) are tabulated in Table 4–4. They used radioactive tracers to determine

TABLE 4–2

RETENTION TIME OF ANALOGS OF PURINE NUCLEOTIDES[a]

Tubercidin monophosphate	11
Tubercidin diphosphate	32
Tubercidin triphosphate	58
2-Fluoroadenosine monophosphate	31
2-Fluoroadenosine diphosphate	47
2-Fluoroadenosine triphosphate	70
Formycin A monophosphate	13
Formycin A diphosphate	31
Formycin A triphosphate	51
Thioguanosine	9
Thioguanosine monophosphate	35
Thioguanosine diphosphate	58
Thioguanosine triphosphate	84
Selenoguanosine	10
Selenoguanosine monophosphate	37
Selenoguanosine diphosphate	60
Selenoguanosine triphosphate	88
Mercaptopurine riboside monophosphate	29
Methylmercaptopurine riboside monophosphate	29
Toyocomycin monophosphate	—
Toyocomycin diphosphate	—
Toyocomycin triphosphate	73
8-Azaadenosine monophosphate	22
Uric acid	9
Thiouric acid	45
Diguanosine triphosphate	83
Diguanosine tetraphosphate	97
Adenosine tetraphosphate	73

[a]*Instrument*: Varian LCS 1000. *Column*: 1 mm × 3 m; packed with pellicular anion exchange resin. *Eluents*: 0.015 M KH$_2$PO$_4$; 0.25 M KH$_2$PO$_4$ in 2.2 M KCl. *Flow rates*: 12 ml/hr; 6 ml/hr. *Starting volume of low conc. Eluent*: 50 ml. *UV* output; 0.08. [Unpublished data, P. R. Brown.]

the elution position of a small number of compounds, mainly amino acids. For about 13 of the peaks, positive identification was made by the conventional techniques of cochromatography plus mass spectrometry, nuclear magnetic resonance, optical spectrometry, and chemical testing of fractions that were collected.

TABLE 4–3
Separation of Various Nucleosides
and Bases by Cation Exchange[a]

Order of elution	Position of maximum (ml)
Pseudouridine	5.5
Uridine	6.2
Ribothymidine	6.6
3-Methyl Uridine	6.8
Uracil	8.2
Inosine	9.2
1-Me inosine	9.2
4-Thiouridine	10.5
4-Thiouracil	10.8
Thymine	10.5
Hypoxanthine	12.2
Guanosine	12.4
N^2-Me guanosine	15.2
N^2-Me guanosine	17.6
Adenosine	18
Bis (4-thiouridine) disulfide	20
N-Me adenosine	22.3
Guanine	24.3
Cytidine	24.4
7-Me guanine	35
N^4-Me_2 adenosine	42.5
Adenine	62
Cytosine	67
1-Me adenosine	72
7-Me inosine	73
3-Me cytidine	72.5
7-Me guanosine	104
1,7-Me_2 guanosine	125

[a] *Instrument*: noncommercial. *Column*: 19.0 × 0.6 cm; packed with BioRad A-6. *Eluent*: 0.4 *M* ammonium formate. *Temperature*: 50°C. *Flow rate*: 0.40 ml/min. [Contributed by Uziel *et al.* (1968).]

It was noted earlier (Volkin and Cohn, 1954; Sober and Peterson, 1957) that in liquid chromatography the order of elution does not follow pK_a values. Brown (1969, unpublished data) in separating barbiturates noted that phenobarbital and mephobarbital have the same retention time even though the phenobarbital has an α-hydrogen on a nitrogen atom and the mephobarbital has a methyl group instead of the hydrogen. Moreover, Katz and

TABLE 4–4

REFERENCE COMPOUNDS FOR URINE
CHROMATOGRAMS[a]

Compound	Elution volume
Adrenaline	14
Argenine	14
Cyanobalamin	14
Cystine	14
Dopamine	14
Histidine	14
Lysine	14
Metamerphrine	14
Noradrenaline	14
Thianine	14
Seratonin	14
Asparagine	18
Citrulline	18
Creatine	18
Glutamine	18
Glucose	18
Creatinine	19
Cytosine	20
Alanine	21
Ergothioneine	21
Methionine	21
Proline	21
Urea	21
Cytidine	21
Phenylalanine	27
Trytanine	27
Pseudouridine	29
Uridine	38
Deoxyuridine	39
Theobromine	39
Uracil	42
Nicotinamide	42
Thymidine	45
Thrasine	49
Caffeine	50

[a] *Instrument*: Varian LCS 1000. *Column*:
0.24 × 100 cm; packed with Aminex BRX
(12–15 μm). *Eluents*: linear acetate gradient
varying in concentration from 0.015 M to
6.0 M (pH 4.4). *Temperature*: ambient. *Flow
rate*: 8 ml/hr: 4 ml/hr. *Sample*: 0.2 ml human
urine. [Contributed by Katz and Burtis
(1969).]

Burtis (1969), using anion exchange chromatography, found that elution positions are a function of chemical structure. Among the relationships are:

1. Functional groups affect elution position most strongly in acidic compounds. The position and structure of these groups also influence the retention time.
2. Functional groups do not affect basic or neutral compounds which elute without retention.
3. Functional groups do contribute to the separation within families of purines and pyrimidines and their nucleosides.
4. Nucleosides differ only slightly from their bases.
5. Amino acids elute earlier than most acidic compounds.

TABLE 4–5

ELUTION VOLUMES OF SOME PURINE DERIVATIVES AND RELATED COMPOUNDS[a]

$$N\underset{1}{\overset{6}{C}}\overset{H}{\underset{5}{C}}\overset{H}{\underset{7}{N}}CH$$

Primary structure:

Compound	Additions to purine structure	Elution Volume (ml)
Theobromine	1—H, 2=O, 3—CH_3, 6=O, 7—CH_3	39
Caffeine	1—CH_3, 2=O, 3—CH_3, 6=O, 7—CH_3	50
Deoxyinosine	1—H, 6=O, 9-(2-deoxy-p-D-ribofuranosyl)	62
Inosine	6=O, 9-p-D-ribofuranosyl	66
Hypoxanthine	3—H, 6=O	75
Deoxyadenosine	6—NH_2, 9-(2-deoxy-p-D-ribofuranosyl)	94
Adenosine	6—NH_2, 9-p-D-ribofuranosyl	101
I-Methylguanine	1—CH_3, 2—NH_2, 3—H, 6=O	118
7-Methylguanine	2—NH_2, 3⁻ H, 6=O, 7—CH_3	120
Theophylline	1—CH_3, 2=O, 3—CH_3, 6=O	137
Adenine	6—NH_2	152
Xanthosine	1—H, 2=O, 3—H, 6=O, 9-p-D-ribofuranosyl	163
Xanthine	1—H, 2=O, 3—H, 6=O	172
Guanosine	1—H, 2—NH_2, 6=O, 9-p-D-ribofuranosyl	214
Deoxyguanosine	1—H, 2—NH_2, 6=O, 9-(2-deoxy-p-D-ribofuranosyl	217
6-Methylaminopurine	6—$NHCH_3$	240
Guanine	2 NH_2, 3—H, 6=O	241
Uric acid	1—H, 2=O, 3—H, 6=O, 8=O, 9—H	550

[a] *Instrument*: Varian LCS 1000. *Column*: 0.24 × 100 cm; packed with Aminex BRX (12–15 μm). *Eluents*: linear acetate gradient varying from 0.015 M to 6.0 M (pH 4.4). *Temperature*: ambient. *Flow rates*: 8 ml/hr; 4 ml/hr. [Contributed by Katz and Burtis (1969).]

6. Aromaticity, additional carboxy and hydroxy groups, increase retention times because of increased acidity.
7. Indole and quinaldic derivatives and benzene rings make equal contribution to retention times, whereas imidazole and pyridine rings contribute less.
8. Compounds of the pyrimidine, amino acid, imidazole, purine, pyridine, indole, quinaldic, and benzoic acid families elute in approximately the listed order, although there is some overlapping.

Table 4–5 shows the effect of various constituents on the purine ring when chromatographed under high pressure on an anion exchange resin. With purine and pyrimidine derivatives, the carbonyls with α-hydrogens have longer retention times (Table 4–5). They demonstrated this very strikingly with hypoxanthine which has an elution volume of 75 ml, xanthine, 172 ml, and uric acid, 550 ml. Using a different chromatography system, Brown (1970) found this order to be the same for the nucleotides of hypoxanthine and xanthine. If, however, an amino group is in the 2 or 6 position, as a rule, the retention times are generally longer than their carbonyl analogs. With a methylamine group, the retention times are even longer.

Another possible way of identifying an unknown peak is to separate several series of homologs under two completely different sets of conditions. The log of the retention volume under one set of conditions is plotted against the retention volumes under the other set. If the unknown compound has a similar structure to one series of homologs, it will fall on or near that straight line. This technique is frequently used in gas chromatography to characterize a peak.

II. Use of Internal Standards

There are several ways in which internal standards may be used. In one, the internal standard is added to the solution before the chromatography process. This method will compensate for errors made in the preparation of the solution. On the other hand, the internal standard may be injected separately, immediately before or after the injection of the sample. In gas chromatography, the requirements for the use of an internal standard in quantitation are stringent. The internal standards must be completely resolved from all unknowns, yet must elute near the peak of interest. It must be similar in concentration and close chemically, but not identical to the unknown. However, in liquid chromatography, it is also possible to add to the solution or inject separately a known quantity of a compound that is thought

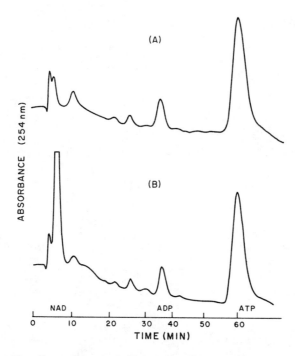

Fig. 4–5. Use of internal standard with human erythrocytes. *Instrument*: Varian LCS 1000. *Column*: 1mm × 3 m; packed with pellicular anion exchange resin. *Eluents*: 0.015 *M* KH$_2$PO$_4$, pH 4.5; 0.25 *M* KH$_2$PO$_4$ in 2.2 *M* KCl, pH 4.5. *Flow rates*: 12 ml/hr; 6 ml/hr. *Starting volume*: 50 ml. *Temperature*: 75°C. *Samples*: (A) nucleotide extract of human erythrocytes; (B) nucleotide extract of human erythrocytes plus NAD. [P. R. Brown, unpublished data.]

to be present. Therefore, this peak will increase in size by a given amount. This procedure has been found to be practical in characterizing peaks of close retention time, especially when several of the possible known compounds are added in consecutive analyses. An example of this method is shown in Fig. 4–5, in which NAD was injected right after a cell extract of human erythrocytes. A large increase in the second peak is noted in the lower chromatogram. The use of an internal standard near an unknown peak which is similar but not identical in concentration and chemical nature is demonstrated in Fig. 4–6. The top chromatogram is a nucleotide extract of *Schistosoma mansoni*, the middle one of the schistosomes plus AMP, and the bottom one has an added amount of GMP. The peak of interest was at 25 minutes. These chromatograms showed that the unknown peak was neither AMP nor GMP. From the use of other internal standards, it was tentatively identified as UDPG or one of the other UDP sugars, all of which have the same retention time under the operating conditions used. Another example of using internal standards is demonstrated in Fig. 4–7. In order to

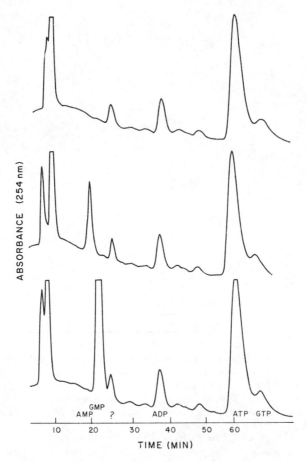

Fig. 4–6. Use of internal standards with *Schistosoma mansoni*. *Instrument*: Varian LCS 1000. *Column*: 1 mm × 3 m; packed with pellicular anion exchange resin. *Eluents*: 0.015 *M* KH$_2$PO$_4$; 0.25 *M* KH$_2$PO$_4$ in 2.2 *M* KCl. *Flow rates*: 12 ml/hr; 6 ml/hr. *Starting volume*: 50 ml. *Temperature*: 75°C. *Sample*: nucleotide extract of *Schistosoma mansoni*, alone and with internal standards of AMP and GMP. [Courtesy of Senft and Brown (unpublished data).]

locate the uracil nucleotides in a cell extract of murine leukemia cells, these nucleotides were added to a cell extract of murine leukemia cells. The upper chromatogram is that of the cells. The same extract to which the uracil nucleotides were added is shown in the lower one. It should be emphasized that this is not always conclusive identification and other means of character-izing these peaks should also be used in conjunction with this use of retention data and internal standards.

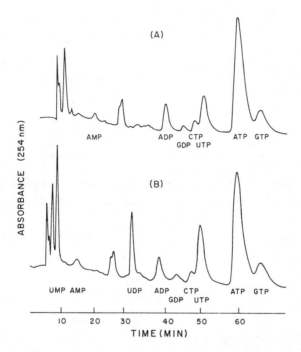

Fig. 4–7. Use of internal standards with murine leukemia cells. (A) Separation of nucleotides in a TCA extract of tissue culture of murine leukemia cells (L5178Y). *Instrument*: Varian LCS 1000. *Starting volume*: 50 ml. *Flow rates*: 12 and 6 ml/hr. *Eluents*: 0.015 M KH$_2$PO$_4$; 0.25 M KH$_2$PO$_4$ in 2.2 M KCl. *Sample*: 4 μl. *UV output*: 0.04 AU. (B) Identification of uridine peaks by the addition of uridine nucleotides to a TCA extract of a tissue culture of murine leukemia cells (L5178Y). *Starting volume*: 50 ml. *Flow rates*: 12 and 6 ml/hr. *Eluents*: 0.015 M KH$_2$PO$_4$; 0.25 M KH$_2$PO$_4$ in 2.2 M KCl. *Samples*: 4 μl cell extract; 1 μl solution of uridine nucleotides (mono-, di-, and triphosphate). *UV output*: 0.04 AU. [Contributed by Brown (1970).]

III. Isotopic Labeling

A variation of the use of standards is the method in which predetermined quantities of a standard radioactive compound are added to the solution. The fractions are collected and by plotting the counts of the fractions, the peak of interest can be identified. This method is especially useful in following cell metabolism of purine and pyrimidine analogs. A plot of the nucleotides in a cell extract of schistosomes containing ^{14}C adenine and guanine nucleotides is shown in Fig. 4–8.

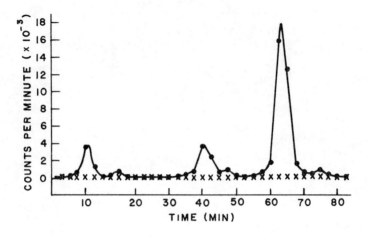

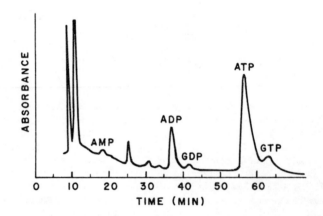

Fig. 4–8. Use of isotopes. *Instrument*: Varian LCS 1000. *Column*: 1 mm × 3 m; packed with pellicular anion exchange resin. *Eluents*: 0.015 *M* KH$_2$PO$_4$; 0.25 *M* KH$_2$PO$_4$ in 2.2 KCl. *Flow rates*: 12 ml/hr; 6 ml/hr. *Starting volume*: 50 ml. *Temperature*: 75°C. [Contributed by Brown (1970).]

IV. Collection and Characterization of Eluent Peaks

Another method of identifying peaks is the collection of fractions and the characterization of the eluent peaks. Characterization methods can be placed in two categories: nonchromatographic identification and chromatographic techniques, either using other than high pressure liquid chromatography or utilizing high performance liquid chromatography with a different column and/or mobile phase.

Since the amount of solution in the fractions is very small, the samples must be concentrated to a suitable dilution. If the eluents are in salt solution, usually the salt concentrations are high and must be removed to obtain usable samples. Also, if the solute is dissolved in an organic solvent, the solvent must be removed. Since the amount of solute present in the eluent peak is so small, it is sometimes necessary to pool the fractions of many analyses of the same sample in order to obtain a large enough amount of the compound to analyze in other ways. Classical column chromatography can be run concurrently using analogous eluents and column packings. Since a large mount of sample is required in this technique, the fraction sizes will be proportionally larger. Therefore, more tests can be run on each fraction to verify the peak identity. Preparative high pressure liquid chromatographs are now available, however, so that larger samples can be obtained in a short time.

There are three general types of identification by nonchromatographic methods; classical microchemical tests, derivative formation, and identification by auxiliary instrumentation. Under the heading of classical microchemical tests, characterization is possible by molecular weight, boiling point, melting point, elemental analysis, or by known colorimetric tests for functional groups. With many organic compounds, derivatives can be formed which can be positively identified. By using auxiliary instrumentation such as infrared, ultraviolet, visible, or fluorescence spectroscopy, mass spectrometry, nuclear magnetic resonance, electron spin resonance, coulometry, polarography, or flame photometry, additional information can be obtained to aid in the positive identification of the compound.

Often it is possible to use other types of chromatographic identification such as thin layer chromatography, paper chromatography, or electrophoresis. If thin layer or paper chromatography is used, several different elution systems may be used to characterize a compound and compare retention values with those in the literature. However, it is often necessary to use one or more of the methods outlined above to identify positively the eluent peaks.

V. Enzymic Peak-Shift

A method of great value in biochemistry in verifying peak identities is the enzymic peak-shift technique. The specificity of enzyme reactions with a nucleotide or class of nucleotides is utilized. This technique is especially useful in the characterization of nucleotides of cell extracts because not only is the identity of the reactant verified, but also that of the product formed. Moreover, these reactions are helpful in clarifying or unmasking a chromatogram. If one nucleotide is present in a large quantity, as, for example, the concentration of ATP in human erythrocytes, it may hide the presence of a small quantity of another nucleotide which has a similar retention time. It is then possible to show conclusively which nucleotides are present. The enzymic peak-shift method is also helpful in the quantitation of a hidden peak or the determination of the shape of a peak which otherwise can be seen only as a shoulder. One requirement for these reactions is that the reagents be readily available, adequately pure, and relatively inexpensive. If any of the added substrates or conversion products absorb in the UV at 254 mm, they must have different retention times than the reactant. An example of an enzymic peak-shift is the use of hexokinase and an excess of glucose to identify the ATP and ADP peaks. ATP + glucose in the presence of hexokinase and Mg^{2+} gives ADP + glucose-6-phosphate.

$$ATP + glucose \xrightarrow[Mg^{2+}]{HK} ADP + glucose\text{-}6\text{-}phosphate$$

In this reaction the ADP increased proportionally in size to the disappearance of the ATP peak. The AMP and guanine nucleotide peaks were not

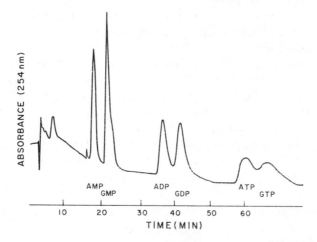

Fig. 4–9. Separation of a TCA extract of adenine and guanine nucleotides. *Instrument*: Varian LCS 1000. *Column*: 1 mm × 3 m; packed with pellicular anion exchange resin. *Starting volume*: 50 ml. *Flow rates*: 12 and 6 ml/hr. *Eluents*: 0.015 M KH_2PO_4 and 0.25 M KH_2PO_4 in 2.2 M KCl. *Sample*: 10 μl of a TCA extract of solutions of AMP, ADP, and ATP and of GMP, GDP, and GTP. *UV output*: 0.32 AU (I-mV recorder). [Contributed by Brown (1970).]

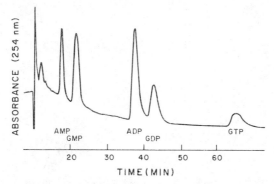

Fig. 4–10. Enzymic peak-shift caused by reaction with hexokinase and glucose. *Instrument*: Varian LCS 1000. *Column*: 1 mm × 3 m; packed with pellicular anion exchange resin. *Starting volume*: 50 ml. *Flow rates*: 12 and 6 ml/hr. *Eluents*: 0.015 M KH_2PO_4 and 0.25 M KH_2PO_4 in 2.2 M KCl. *Sample*: 10 μl of a TCA extract of reaction mixture. *UV output*: 0.32 AU (I-mV recorder). [Contributed by Brown (1970).]

altered. Therefore, it can be seen that positive identification is established for both the ATP and ADP peaks. Chromatograms of a standard solution of adenine and guanine nucleotides before the reaction with hexokinase is shown in Fig. 4–9. After the reaction, no ATP can be seen and the ADP peak is doubled (Fig. 4–10). Another typical enzymic peak-shift is the reaction of pyruvate kinase and an excess of phospholenolpyruvate (PEP) in the presence of magnesium chloride with diphosphate nucleotides.

$$\text{ADP} + \text{PEP} \xrightarrow[\text{Mg}^{2+}]{\text{PK}} \text{ATP} + \text{pyruvate}$$

$$\text{GDP} + \text{PEP} \xrightarrow[\text{Mg}^{2+}]{} \text{GTP} + \text{pyruvate}$$

The diphosphates are phosphorylated to 5'-triphosphate nucleotides after the cell extract has been incubated with pyruvate kinase and PEP for 10 minutes. As can be seen from Fig. 4–11, the diphosphate nucleotide peaks have disappeared completely and the triphosphate nucleotide peaks have increased in size. The peak which has a retention time of 30 minutes is PEP. In all enzymic peakshift methods, after treatment with the enzyme the solutions were treated with trichloracetic acid (TCA) to precipitate any protein or acid-insoluble material, thus preventing accumulation of the enzyme on the column. The nucleotides were not affected when the TCA was removed by extraction with water-saturated diethyl ether.

Many other enzyme reactions can be used in the enzymic peak-shift method of characterizing nucleotides. Almost any of the phosphotransferases in the groups numbered 2.7.1–2.7.4 by the Enzyme Commission can be used with an excess of the appropriate substrate. In the presence

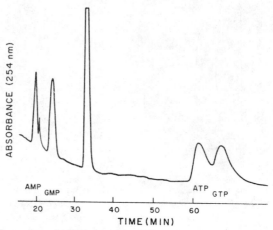

Fig. 4–11. Enzymic peak-shift caused by PEP and pyruvate kinase. *Instrument*: Varian LCS 1000. *Column*: 1 mm × 3 m; packed with pellicular anion exchange resin. *Starting volume*: 50 ml. *Flow rates*: 12 and 6 ml/hr. *Eluents*: 0.015 M KH$_2$PO$_4$; 0.25 M KH$_2$PO$_4$ in 2.2 M KCl. *Sample*: 10 μl of a TCA extract of reaction mixture. *UV output*: 0.32 AU (I-mV recorder). [Contributed by Brown (1970).]

of an enzyme from group 2.7.1, a phosphate group is transferred to a sugar molecule or other alcohol acceptor, resulting in the formation of ADP and a phosphorylated alcohol. Examples of these enzymes are fructokinase, choline kinase, or pantetheine kinase. The enzymes hexokinase and pyruvate kinase, which were discussed previously, belong in this category.

$$\text{ATP} + \text{D-fructose} \rightleftharpoons \text{ADP} + \text{D-fructose-6-phosphate}$$
$$\text{ATP} + \text{choline} \rightleftharpoons \text{ADP} + \text{choline phosphate}$$
$$\text{ATP} + \text{pantetheine} \rightleftharpoons \text{ADP} + \text{pantetheine 4'-phosphate}$$

With enzymes in the group 2.7.2, a carboxyl group is the acceptor. Examples of these enzymes are aspartate kinase and acetate phosphotransferase.

$$\text{ATP} + \text{L-aspartate} \rightleftharpoons \text{ADP} + \text{4-phospho-L-aspartate}$$
$$\text{ATP} + \text{acetate} \rightleftharpoons \text{ADP} + \text{acetyl phosphate}$$

Nitrogenous groups act as acceptors with enzymes from group 2.7.3 and a well-known example is the reaction of ATP and creatine in the presence of creatine kinase to form ADP and phosphocreatine. When phospho groups are acceptors, the enzymes are classified in group 2.7.4. Well-known examples of this group are adenylate kinase, guanylate kinase, and GTP-adenylate kinase. With adenylate kinase, a phosphate group is transferred from ATP to AMP, producing 2 moles of ADP

$$\text{ATP} + \text{AMP} \rightleftharpoons 2\text{ADP}$$

With guanylic kinase, the acceptor is GMP and with GTP-adenylate kinase, the donor is GTP and the acceptor AMP.

$$ATP + GMP \rightleftharpoons ADP + GDP$$
$$GTP + AMP \rightleftharpoons ADP + GDP$$

Pyrophosphotransferases of the group 2.7.6 can also be used to identify ATP and AMP. With both ribophosphate pyrophosphokinase, and thiamine pyrophosphokinase, a pyrophosphate group is transferred to the substrate, with the resultant formation of AMP and the pyrophosphorylated substrate.

Certain nucleotidyltransferases of the group 2.7.7, such as UDPG pyrophosphorylase, can serve to identify both UTP and UDPG. Ethanolaminephosphate cytidylyltransferase can identify CTP.

$$UTP + D\text{-glucose} \rightleftharpoons pyrophosphate + UDPG$$
$$CTP + ethanolamine\ phosphate \rightleftharpoons pyrophosphate + CDP$$
$$ethanolamine$$

The reaction GDP and mannose-l-phosphate in the presence of GDP-mannose phosphorylase can be used to characterize GDP.

$$GDP + D\text{-mannose l-phosphate} \rightleftharpoons orthophosphate + GDP\text{-mannose}$$

Enzymes of the group 3.6.1 which act on the anhydride bonds can also be used. Examples of these enzymes is ATPase (3.6.1.3). With H_2O the triphosphate nucleotides react to form the diphosphates and orthophosphate. Diphosphates as a class can be identified by the use of nucleoside diphosphatase. In the presence of this enzyme, nucleoside diphosphates react with water to form a monophosphate nucleotide and orthophosphate. Deoxy-CTPase (3.6.1b) can be used to identify specifically dCTP and dCMP.

$$dCTP + H_2O = dCMP + pyrophosphate$$

Enzymes in the classification 3.5.4, such as adenosine demaminase (3.5.4.4), cytidine deaminase (3.5.4.5), AMP deaminase (3.5.4.6), deoxy-CMP deaminase (3.5.4), and ADP deaminase (3.5.4.7) are specific in their deamination reactions and can therefore be used to characterize adenosine, cytidine, AMP, dCMP, and ADP, respectively.

$$Adenosine + H_2O \rightleftharpoons inosine + NH_3$$
$$Cytidine + H_2O \rightleftharpoons uridine + NH_3$$
$$dCMP + H_2O \rightleftharpoons dUMP + nH_3$$
$$ADP + H_2O \rightleftharpoons IDP + NH_3$$

With the enzyme UDPG-dehydrogenase (1.1.1.22) not only can the reactant UDPG and NAD be identified, but also the products UDPGA and NADH.

$$UDPG + NAD = UDPGA + NADH$$

When a reaction is reversible, this reaction can be coupled to a second enzymic reaction, either to remove a product of the first reaction or to drive the first to completion. If the second reaction is readily reversible, its equilibrium can be shifted to the right by adding an excess of substrate. An example of a coupled enzymic reaction is the reaction of ADP in the presence of adenylate kinase (AK) to form AMP and ATP. This reaction can be driven to completion in presence of AMP deaminase. AMP deaminase converts AMP to IMP and NH_3. Therefore the products of their reaction are ATP, IMP, and NH_3.

$$2 ADP \xrightarrow[AMP]{AK} AMP + ATP$$

$$AMP \xrightarrow[\text{deaminase}]{} IMP + NH_3$$

$$2 ADP \xrightarrow[\substack{AMP \\ \text{deaminase}}]{AK} ATP + IMP + NH_3$$

Another coupled reaction is the use of hexokinase and glucose in conjunction with the adenylate kinase reaction. The ATP reacts with the glucose until no ATP remains; thus, AMP and glucose-6-phosphate accumulates.

$$2 ADP \xrightarrow{AK} ATP + AMP$$

$$ATP + \text{glucose} \xrightarrow{HK} ADP + \text{G-6-P}$$

$$ADP + \text{glucose} \xrightarrow[HK]{AK} AMP + \text{G-6-P}$$

Since only AMP absorbs in the UV at 254 nm, it is possible to watch the formation of a large AMP peak. In this case, and the case previously noted, the reactions are specific for the disappearance of ADP. In the former, however, two nucleotides accumulate, ATP and IMP, while in the latter only one is produced, AMP. Another example of driving a reversible reaction to completion is the coupling of the adenylate kinase reaction to that of the creatine kinase reaction. In this reaction, AMP and ATP are converted by the adenylate kinase to ADP. In the presence of creatine kinase and an excess of phosphocreatine, the reaction continues until there is no AMP or ADP in the reaction mixture. Thus, only phosphocreatine and creatine, neither of which absorbs in the UV at 254 nm, and ATP remain.

$$AMP + ATP \xrightarrow{AK} 2 ADP$$

$$2 ADP + 2 \text{ phosphocreatine} \xrightarrow{CK} 2 ATP + 2 \text{ creatine}$$

$$AMP + 2 \text{ phosphocreatine} \xrightarrow[CK]{AK} ATP + 2 \text{ creatine}$$

Although other coupled reactions for enzymic peak-shifts are possible, the reactions outlined are especially useful because of the high specificity of these enzymes for adenine nucleotides. In most cellular extracts, the concentrations of the adenine nucleotides present are much greater than those of other nucleotides. Therefore, the enzymic peak-shift method for the adenine nucleotides is of special value in unmasking peaks of the non-adenine compounds ordinarily hidden by the larger adenine nucleotide peaks.

The enzymic peak-shift method can be used to identify the peaks of compounds for which no standard is available. For example, when human red blood cells are incubated with either 2-fluoroadenine or 2-fluoroadeno-sine, compounds are formed which have peaks with longer retention times than ADP and ATP. It was assumed that the peaks were those of the 5'-diphosphate of 2-fluoroadenosine (2-FADP) and the 5'-triphosphate of 2-fluoroadenosine (2-FATP). Therefore, the cell extract containing the 2-FATP was reacted with hexokinase and glucose. It was found that both ATP and the 2-FATP shifted completely to the ADP and to the peak tentatively identified as 2-FADP. This shift is shown in Fig. 4–12. Some enzymes that

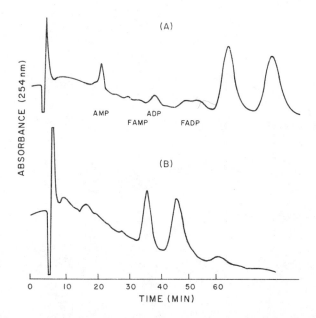

Fig. 4–12. Enzymic peak-shift of 2-FATP by hexokinase. (A) TCA extract of human erythrocytes that had been incubated with 2-fluoroadenosine. (B) TCA extract of same incubation mixture that had been treated with hexokinase and glucose. *Instrument*: Varian LCS 1000. *Column*: 1 mm × 3 m; packed with pellicular anion exchange resin. *Eluents*: 0.015 M KH$_2$PO$_4$, pH 4.5; 0.25 M KH$_2$PO$_4$ in 2.2 M KCl, pH 4.5. *Flow rates*: 12 ml/hr; 6 ml/hr. *UV output*: 0.04 AU. *Temperature*: 75°C. [Brown and Parks (unpublished data).]

can be used in the enzymic peak-shift method to identify peaks in chromatograms of cell extracts are listed in Table 4–6.

Isotopes can also be used in conjunction with the enzymic peak-shift technique. A good example of combining several techniques to identify peaks was demonstrated by Nelson and Parks (1972) in their study on the

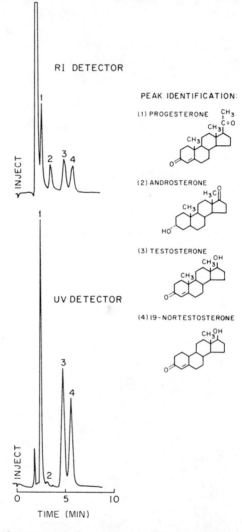

Fig. 4–13. Steroids. *Instrument*: du Pont 820 liquid chromatograph equipped with both UV and RI detectors. *Column*: 2.1 mm i.d. × 1 m; packed with 1% BOP on Zipax. *Eluent*: heptane. *Flow rate*: 1 cm³/min; 600 psig. *UV output*: 0.32 AUFS, RI 8 × 10⁻⁵ RIFS. *Temperature*: ambient. [Contributed by Henry *et al.* (1971b).]

synergistic effect of thioguanine and 6-methylmercaptopurine riboside on sarcoma 180 cells. They compared the retention times obtained by high pressure liquid chromatography of unknown peaks to those of authentic samples. They used TG-^{35}S in isotopic labeling experiments and found that the peak of a radioactive compound with a retention time of 35 minutes corresponded to that of a known sample of TGMP (Fig. 3–22). Using thin

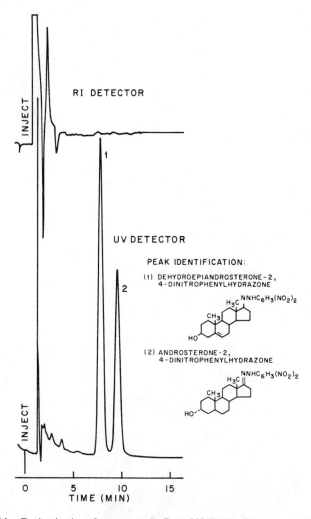

Fig. 4–14. Derivatization. *Instrument*: du Pont 820 liquid chromatograph. *Column*: 2.1 mm i.d. × 1 m; 1% BOP on Zipax. *Eluent*: heptane. *Pressure*: 1 cm³/min; 600 psig. *Temperature*: Ambient. *UV output*: 0.32 AUFS, RI 8 − 10⁻⁵ RIFS. [Contributed by Henry *et al.* (1971b).]

TABLE 4-6

ENZYMES THAT CAN BE USED IN THE "ENZYMIC PEAK-SHIFT" METHOD OF IDENTIFYING PEAKS IN CELL EXTRACT CHROMATOGRAMS

Trivial name	EC Number	Systematic name
UDPG-dehydrogenase	1.1.1.22	UDPG–glucose; NAD oxidoreductase
Hexokinase	2.7.1.1	ATP: D-hexose 6-phosphotransferase
Fructokinase	2.7.1.4	ATP: D-fructose 6-phosphotransferase
Choline kinase	2.7.1.32	ATP: pantothenate 4'-phosphotransferase
Pantotheine kinase	2.7.1.33	ATP: pantothenate 4'-phosphotransferase
Pyruvate kinase	2.7.1.40	ATP: pyruvate phosphotransferase
Aspartate kinase	2.7.2.4	ATP: L-aspartate 4-phosphotransferase
Acetate phosphotransferase	2.7.2.1	ATP: acetate phosphotransferase
Carbonate kinase	2.7.2.2	ATP: carbonate phosphotransferase
Creatine kinase	2.7.3.2	ATP: creatine phosphotransferase
Adenylate kinase	2.7.4.a	ATP: AMP phosphotransferase
Guanylate kinase	2.7.4.b	ATP: GMP phosphotransferase
GTP-adenylate kinase	2.7.4.d	GTP: AMP phosphotransferase
Ribosphosphate pyrophosphate kinase	2.7.6.1	ATP: D-ribose-5-phosphate pyrophosphotransferase
Thionine pyrophosphokinase	2.7.6.2	ATP: thionine pyrophosphotransferase
UDPG-pyrophosphorylase	2.7.7.9	UTP: O-glucose-l-phosphate uridylyltransferase
Ethanolaminephosphate cytidylyltransferase	2.7.7.14	CTP: ethanolaminephosphate cytidylyltransferase
GDP-mannose phosphorylase	2.7.7.e	GDP: D-mannose-l-phosphate guanylyltransferase
ATPase	3.6.1.3	ATP-phosphohydrolase
Deoxy-CTPase	3.6.1.b	Deoxy-CTP nucleotideoxyhydrolase
Adenosine deaminase	3.5.4.4	Adenosine aminohydrolase
Cytidine deaminase	3.5.4.5	Cytidine aminohydrolase
AMP-deaminase	3.5.4.6	AMP-aminohydrolase
Deoxy-CMP deaminase	3.5.4	Deoxy-CMP aminohydrolase
ADP-deaminase	3.5.4.7	ADP-aminohydrolase
IMP-cyclohydrolase	3.5.4.10	IMP-1,2-hydrolase
Succinic thiokinase	6.2.14	Succinic: CoA ligase

layer chromatography, they cochromatographed, using three different eluent systems, the radioactivity in the peaks with a standard sample of TGMP. As a final proof of the identity of the TGMP, they used the enzymic peak-shift technique and shifted the TGMP to TGTP in the presence of guanylic kinase and nucleoside diphosphokinase.

VI. Derivatization

A method that has been widely used in gas chromatography is the formation of a derivative of a compound. This can also be used in liquid chromatography and can be utilized in much the same way as the enzymic peak-shift method to identify peaks. Ideally, the peak of the reactant will disappear and a new peak of the product formed in the reaction will appear. This peak-shift will serve not only to identify product and reactant, but also to clarify or "unmask" the chromatogram in certain areas, if needed, as in the enzymic peak-shift method. The major considerations for selecting a reactant for derivatization, is that the reaction must be complete, rapid, and specific for a compound or a functional group. It is preferable that both the substrate and the product can be characterized by the detector used. However, it is possible to use derivatization so that a compound which does not absorb in the UV can be converted to one that absorbs strongly. In this way sensitivity can be increased. Also, derivatives of compounds can be separated even though the free compounds cannot be resolved. An excellent example of a use of derivatives is illustrated by Henry *et al.* (1971b) who used 2,4-diphenylhydrazone derivatives to separate many of the major types of steroids, (Figs. 4–13 and 4–14).

It may be necessary to use one or more of the methods outlined above to identify positively the eluent peaks. In the final analysis, the researcher must make his own judgment as to what comprises identification of an eluent peak. This will depend on the type of sample, availability of comparable standard compounds, and the references in the literature.

CHAPTER 5

QUANTITATION

In order to perform quantitative analyses or kinetic studies by high pressure liquid chromatography, it is necessary to be able to determine accurately the concentration of one or more components present. With the measurement and calculation methods available, high pressure liquid chromatography is a highly accurate analytical technique, comparable in accuracy to gas chromatography. The quantitative results obtained depend on the complete process of analysis from preparation of the sample to interpretation of the results. As in a chain, the analysis is no better than the weakest link. Therefore, each step must be considered in developing techniques and interpreting the final results.

There are four basic steps in quantitative liquid chromatography: (1) chromatography, (2) integration, (3) determination of sample composition, and (4) statistical analysis of the data. In the chromatography step, an analog signal is generated by the detector and recorded in the form of chromatographic peaks. The area under the peak is converted to digital data in the integration process. Peak areas can be integrated either by manual methods or by using integrating devices. The calculation step consists of relating these data to the composition of the sample. In the final step, the data are expressed through statistics.

I. Chromatography

A. SAMPLE PREPARATION

Using high pressure liquid chromatography in biochemical and bio-medical research, one of the most important parts of the analytical procedure is the preparation of the sample. If the samples are not properly and quickly handled, compounds such as nucleotides can be easily degraded by the enzymes present in tissues. Pipetting precision is essential because large errors in accuracy can be caused by relatively small pipetting errors. In preparing cell extracts using the diethyl ether extraction of TCA from a sample, volume changes can cause as much as 10% difference in reproducibility. The presence of a large concentration of salts also causes differences in chromatograms which result in quantitative errors. In making up solutions of samples or standards, the careful calculation of molarity or concentration is necessary. Too concentrated a solution may cause overloading of the column; thus poor peak shapes are obtained, or there will be problems in the subsequent chromatograms. Also errors in the final data can result from faulty weighings or dilutions of standards, reagents, or samples. Not only is sample preparation important, but care must be taken in sample storage because problems such as changes in concentration due to evaporation, contamination, and salting-out effects can prevent accurate analyses. In addition, some nucleotides may catalyze their own decomposition if not stored under appropriate pH conditions.

It is of the utmost importance that not only the procedure for preparation and handling of the sample be standardized, but also the standard or references should be handled and stored in the very same way as the samples. The probability of accurate results will be increased if extreme care is taken in the standardization procedure.

B. SAMPLE INTRODUCTION

Since the sample volume used in high pressure liquid chromatography is so small, it is critically important that the sample be carefully and quantitatively introduced into the column. The syringe must be absolutely clean and not blocked, no air bubbles must be present, and the sample volume must be measured precisely. If possible, the syringe should be rinsed several times with the sample solution so that there is no dilution effect. The needle must be wiped dry to prevent extra drops from being injected. For this purpose, Kimwipes should be used to prevent contamination by a used cloth. In most instruments, the needle fits right into the column so that no sample is lost during the injection process. If the stopped-flow technique of injection is used, the pressure is stopped for the injection process. The injection technique is not as critical using this technique as it is when the sample is injected into the flowing stream through a septum, i.e., pressure is not stopped. It is also important to make sure the syringe is emptied completely. The sample must be injected slowly so that all the liquid is introduced evenly in the column, thus preventing the formation of a pool of sample at the injection port.

C. SEPARATION OF PEAKS

In the chromatography process, in which the compounds migrate differentially, there may be decomposition of components during the chromatographic process. Care must be taken to avoid thermal or chemical degradation of any of the compounds present. Decomposition may lead to the presence of new peaks, the absence of expected peaks, or the masking of peaks normally in the chromatogram. If there is peak overlap or tailing, different eluents, column packings, or change in temperature should be tried. If, however, separations which were previously good deteriorate, it is possible that the column packing has changed. If the problem is due to an accumulation of impurities on the column, the column should be washed well. If, however, washing does not help, it is possible that the packing material has decomposed irreversibly and a new column should be tried. For an example of the poor separation caused by column deterioration see Fig. 3–23. In the upper chromatogram there is excellent separation of ATP and GTP in a cell extract of rat liver. However, these peaks in the same sample are poorly resolved in the lower chromatogram which had been run a few days before on a column that had been used for $2\frac{1}{4}$ years.

One problem that always faces chromatographers is whether there is any retention of solutes on the column that will bleed off in subsequent runs. The process is referred to as "ghosting." In order to determine whether or not there is "ghosting," it is possible to run an experiment using isotopes. An experiment such as this was carried out by Brown (1970). A standard solution of adenine nucleotides was run followed by a cell extract containing ^{14}C adenine and guanine nucleotides. Fractions were collected for both the samples and the low concentrate wash after each sample. Another standard solution of adenine nucleotides was run and fractions of this sample and the subsequent wash were also collected. The radioactivity of all the fractions from each solution was counted and it was found that peaks of the radioactive sample as plotted corresponded to the chromatogram from the analyzer (see Fig. 4–4). The wash solutions and the standard solution run both before and after the ¹⁴C samples were found to have only background levels of radioactivity. Therefore, it was concluded that the nucleotides were eluted completely.

D. DETECTION AND AMPLIFICATION

Important detector specifications are sensitivity, linearity, and specificity. Errors in detection and amplification can be due to lack of familiarity with electronics. Knowledge of detector specificity is important in quantitative analysis. For example, all compounds do not absorb in the UV. Therefore, some compounds will not have peaks in a chromatogram obtained using a UV detector. Also, as was discussed in the section on detectors, column flow rate, detector electronics, and temperature all affect the quantitative results. It is important to know which parameters are critical with the detector being used so that proper precautions can be taken. It is necessary to know if the concentration of the compound to be quantitated is linear with the area of the peak. Modified Beer's laws plots should be determined for each component in a sample. An example of the plot of concentration of AMP (in nanomoles) vs the area (in square centimeters) of the AMP is shown in Fig. 5–1.

II. Integration

The integration step consists of converting the detector signal into numerical data. Since the detector signal is usually in the form of peaks in a chromatogram from a strip chart recorder, it is necessary to convert the

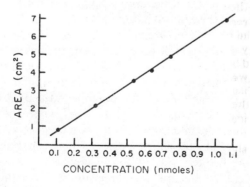

Fig. 5–1. Plot of concentration (in nanomoles) of AMP vs area (cm²) of AMP peak. [Reproduced from Brown (1970).]

peak area or heights to numbers. This can be done manually or by electronic techniques. If electronic techniques are used, neither a recorder nor an operator may be necessary for the conversion step. However, it is useful to have the chromatogram as it is sometimes possible to spot differences in peak areas or shapes that might not be detectable as easily with numbers alone. These differences could be due to actual reactions in the cell or cell extracts or could be symptoms of trouble in the chromatographic system. When a strip chart recorder is used, characteristics that can affect the accuracy of the results are the dead band, linear range, pen speed, shifting of the balance point, chart paper degree of filtering, or irregular pen movements. The first two characteristics are fully described in Varian's "Basic Liquid Chromatography."

A. Manual Methods

The manual methods for converting detector signals into numerical data are: (1) planimetry; (2) measurement of peak height; (3) measurement of height times width at half-height; (4) triangulation; and (5) cut and weigh.

A planimeter is a device which measures area by tracing the perimeter of the peak. (Fig. 5–2). The area is presented digitally on a dial. Since the precision and accuracy of this method depends not only on the device itself but largely on the steadiness and skill of the operator, the peaks should be traced several times and the results averaged. This method is not as accurate as the other methods. It is tedious and time-consuming as well. The second manual method is the measurement of peak height (Fig. 5–3). Although this procedure is simpler than the measurement of peak areas, it is not as accurate since peak height does not always change linearly with sample size. This is especially true if operating conditions vary or when the column is overloaded. In the third method, the area is approximated by multiplying the height of the peak by the width at half-height (Fig. 5–4). The accuracy of this method is affected by the measurement of the width since a narrow peak can adversely affect the precision. Rather than using the base, the width at half-height is used to reduce the errors due to tailing of the peaks and any baseline irregularities. The fourth method is that of triangulation in which a triangle is constructed by drawing tangents to the slope of the peak (Fig. 5–5). The area is calculated by the triangle formula $A = \frac{1}{2}BH$, where the base is taken as the distance between the intersection of the two tangents with baseline and the height is measured by the shortest distance between the baseline and the intersection of the two tangents. With this method, it is assumed that the peak is symmetrical. The accuracy and precision of this method is dependent on the skill of constructing the triangle and the complexity of drawing the tangent lines. A slight error in the placement of the tangents can have a large

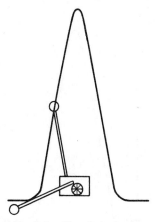

Fig. 5–2. Use of planimeter.

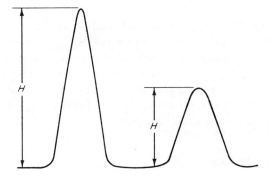

Fig. 5–3. Measurement of peak height.

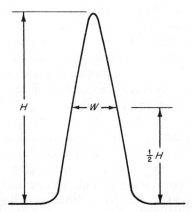

Fig. 5–4. Measurement of area using peak height and width at half-height.

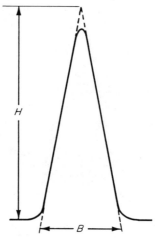

Fig. 5–5.   Triangulation.

effect on the results. The fifth manual method used is the cut-and-weigh method; the peak is cut out of the chromatogram and weighed on an analytical balance. The constancy, weight and moisture content of the chart paper, and the skill in cutting are factors that affect the accuracy. Moreover, the chromatogram is destroyed; this can be a serious disadvantage. It has been found that cutting and weighing of a xerox copy of the chromatogram will minimize these disadvantages.

B. INTEGRATING DEVICES

A number of integrating devices have been designed for direct attachment to recorders so that the peak areas are recorded and integrated simultaneously. Examples of these are electronic ball and disc, and analog and voltage integrating devices. One of the most widely used integrators in gas chromatography is the ball and disc type, manufactured by Disc Instruments Inc., because it provides automation relatively inexpensively. The accuracy and precision obtained is dependent on careful adjustment of the operation of the recorder and is limited by the mechanical performance of the recorder. However, although the results are excellent when the baseline of the chromatogram is steady and the separation of the peaks is good, the disc recorder is difficult to use under other conditions. The electronic digital integrator is one in which the chromatographic input signal is fed into a voltage to frequency converter and an output pulse rate, proportion to the peak area, is generated. The pulses from the voltage to frequency converter are accumulated and printed out as a measure of peak area when the slope detector senses a peak. The major advantages of this integrator are the wide linear range, high count rate, and sensitive power detection. Although they are expensive, the cost may be worthwhile because of their high precision, sensitivity, and the rapidity with which they work.

Computers are being used to integrate peak areas. Although digital integrators are accurate and sensitive devices for measuring the peak areas, they do not handle the calculations or interpret the data. Computer systems that have been used include off-line, hybrid, dedicated computers, or multichannel dedicated, or time- shared computers. In the off-line, batch-processing approach, the data on printed or punched paper tape from the digital integrator is manually transferred to a computer for processing. The hybrid has an integrator on-line to a computer. In the dedicated computer, a single computer is attached to a liquid chromatograph. This type of system is used mainly for research studies. In the multichannel dedicated computer many chromatographs are on-line to one computer dedicated to this work. In the time-shared computer, chromatographs, as well as other analytical instruments, are on-line to one large computer. Time-shared services have made off-line computation available to small laboratories and can provide good computorial powers and storage of such data as compound name, retention time, and response factor.

Burtis and Gere (1970) investigated six integration techniques and tabulated the precision of the results obtained by the various integration methods. The time required to perform the integration varied over a wide range. Ten chromatograms were picked at random for evaluation of these techniques and operators skilled in the various techniques performed the integration. The results (Table 5–1) show that the manual methods are expensive in terms of time expended on the integration. The precision of the digital integration was far better than the majority of the manual methods. The disc ® integrator, although not as fast or as precise as the digital integrator, was more rapid and more precise than manual methods. The digital integrator can correct the peak area for baseline drift and offset, sense and separate shoulders and overlapped peaks, and identify peaks by relative retention time. The on-line computer also has many advantages. This system not only detects peaks by relative retention times, but it also calculates area percentage of each peak, calculates composition results in desired units, and types an output of the quantitative result. Although the time of integration and precision are about as good as the digital integrators, its main advantage is that it has greater manipulative power.

When quantitating nucleotide peaks, at least 10 samples of a standard solution of each nucleotide should be run. Because absorbance is directly related to molar extinction coefficients at the wavelength used, separate calibrations are required for each nucleotide. The concentration of a

TABLE 5-1

Comparison of Integration Methods[a]

	Planimeter		Triangulation		$H \times W$ at $\frac{1}{2}H$		Cut and weigh		Disc		Digital	
	Avg.	o rel (%)	Avg.	o rel (%)	Avg.	o rel (%)	Avg.	o rel (%)	Avg.	o rel (%)	Avg.	o rel (%)
Peak 1	0.04	20	0.05	14	0.04	12	0.005	14	0.04	18	0.03	0.10
Peak 2	4.84	1.65	4.79	8.77	4.52	3.76	5.04	1.98	4.57	0.44	4.56	0.88
Peak 3	14.01	5.64	13.70	4.53	13.56	1.62	14.99	2.80	14.05	1.00	14.06	0.36
Peak 4	18.52	3.29	18.52	4.75	18.66	3.05	18.40	1.22	18.73	2.46	18.73	0.27
Peak 5	8.12	5.67	8.55	3.74	8.56	2.22	8.01	2.12	8.07	2.60	8.24	0.49
Peak 6	20.18	6.49	20.32	3.25	20.07	2.04	20.09	1.39	20.21	1.14	20.16	0.20
Peak 7	16.11	3.66	15.83	1.58	16.41	3.71	15.85	1.45	16.06	0.68	15.95	0.38
Peak 8	18.19	2.03	18.25	1.81	18.17	1.65	17.58	1.19	18.28	0.71	18.27	0.49
Time to trace (min)	45–50		45–60		50–60		100–120		15–30		5–10	
Precision (o rel)[b]	4.06%		4.06%		2.58%		1.74%		1.29%		0.44%	

[a]From Burtis and Gere (1970).
[b]Excludes peak 1.

TABLE 5–2

Tabulation of Areas Under
the Peaks of a Standard
Solution of AMP[a,b]

Spectrum no.	Area/nmole
163	6.42
171	6.81
173	6,74
174	6.87
175	6.92
176	6.49
180	6.76
181	6.81
182	6.52
183	6.69
	Mean 6.71

Standard deviation   0.18
Coefficient of variation 2.8

[a]From Brown (1970).
[b]The AMP solution was 6.08×10^{-5} M. The volume of sample solution injected ranged from 6 to 8 μl. *Instrument*: Varian LCS 1000. *Starting volume*: 50 ml. *Flow rates*: 12 and 6 ml/hr. *Eluents*: 0.015 M KH_2PO_4 and 0.25 M KH_2PO_4 in 2.2 M KCl. *UV output*: 0.08.

nucleotide in a cell extract is then determined by relating the peak area to the peak area of 1 nmole of the standard nucleotide. In all cases, the same method of calculation should be used for standards as for samples. An example of the use of this method of calculating the peak area by multiplying the height times the width at half-height, is the AMP run by Brown (1970) (Table 5–2). The standard deviation of the 10 samples was 0.18 and the coefficient of variation 2.8%. The concentration of all adenine nucleotides and cell extracts can be calculated by relating peak areas to that of the standard AMP since all adenine nucleotides have the same molar extinction coefficient. A comparison of values of total adenine nucleotide content as determined by high pressure liquid chromatography and by an enzymic assay in four different blood samples showed close agreement. These values are shown in Table 5–3 (Brown, 1970).

TABLE 5–3

TOTAL ADENINE NUCLEOTIDES (AMP + ADP + ATP) (nmoles/μl)[a,b]

	Sample			
	H	P	A	S
Varian LCS 1000	0.387	0.344	0.354	0.229
Enzymic analysis	0.393	0.332	0.348	0.230

[a]From Brown (1970).

[b]*Instrument*: Varian LCS 1000. *Column*: 1 mm × 3 m; packed with pellicular anion exchange resin. *Eluents*: 0.015 M KH$_2$PO$_4$; 0.25 M KH$_2$PO$_4$ in 2.2 M KCl. *Flow rates*: 12 ml/hr, 6 ml/hr. *Starting volume of low concentration eluent*: 50 ml. *UV output*: 0.08.

III. Calculation

In liquid chromatography, as in other chromatographic methods, the composition of the solute must be calculated. There are three principal methods used for relating the digital data to the composition of the sample: normalization, internal standardization, and calibration techniques.

A. NORMALIZATION

In the normalization method it is assumed that the entire sample is eluted and detected. The percentage of peak x, in Fig. 5–6, is given by the formula:

$$\%x = \frac{A_x}{A_x + A_y + A_z} \times 100$$

A_y, A_x, and A_z represent the individual peak areas. This method can be useful when analyzing complex mixtures. Since the relative peak areas obtained in liquid chromatography are not always related to composition, because the response of a given detector may be different for each molecular type or class of compound, it is necessary to use response factors. This will

Fig. 5–6.   Normalization.

correct not only for the response of the detector, but also can serve as a factor for band-broadening with time. Since response factors are not easy to estimate, they are best obtained by analyzing standard samples. Response factors relative to a reference compounds, are given by the formula:

$$F_x = \frac{F_r A_r W_x}{A_x W_r}$$

A_x and A_r are the areas of the solute and reference peaks, respectively, W_x and W_r are the amount of the solute and reference compounds, respectively, and F_r is the response factor assigned to the reference compound. Corrected areas are obtained by multiplying the area by the relative response factor. Since different detectors operate on different principles, different factors must be calculated for different detectors.

B. INTERNAL STANDARDS

In order to eliminate apparatus and procedure errors, internal standards can be used. The requirements for an internal standard are as follows:

1. It must be completely resolved from all the unknowns
2. It must elute near the peak of interest
3. It must be similar in concentration to the peak of interest
4. It must be chemically similar but not present in the original sample
5. It must be chemically inert

The percentage of one compound can be calculated by using the formula:

$$\%x = \frac{A_x}{A_{is}} \times \frac{F_x}{F_{is}} \times \frac{W_{is}}{W_s} \times 100$$

The areas are represented by A_x and by A_{is}, the correction action by F_x and F_{is} of the unknown peak and internal standard, respectively, and the weights by W_s and W_{is} of the sample and internal standard, respectively. This procedure is used when the sample is not completely eluted, when it is necessary to measure one or two peaks accurately, or to compensate for errors made during the preparation of the sample. A graphical interpretation of results are shown in Fig. 5–7.

C. CALIBRATION

In calibration, one external standard is used so that there can be direct comparison of the peak area of the sample to that of a standard which has

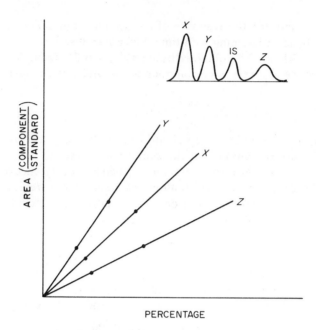

Fig. 5–7. Use of internal standard.

been injected directly. A calibration factor K is determined by injecting the standard solutions and the results are expressed by the formula:

$$\%x = A_x K$$

This technique depends upon the precise injection of the standards. It is particularly useful in analyzing simple mixtures or for trace analysis. If the peak areas are plotted versus the known weight of the compound, this is known as absolute calibration (Fig. 5–8). If, however, the areas of the sample divided by the standard are plotted against the weight of the sample divided by the standard, this is known as a relative calibration curve (Fig. 5–9). In practice, an accurately known amount of an internal standard is added to an unknown sample and the mixture is chromatographed. The area ratios are measured and from the calibration graph, the weight ratios of sample to standard are obtained. Since a known amount of standard was added, the amount of unknown sample can be calculated. The advantages of this technique are that the quantities injected need not be accurately measured and the detector response does not have to be known nor remain constant. The major disadvantage, especially in dealing with nucleotide pools of cell extracts, is the difficulty in finding a standard that does not interfere with the components in the sample.

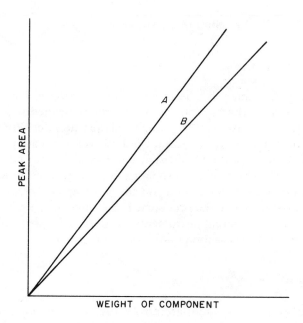

Fig. 5–8.   Absolute calibration curve.

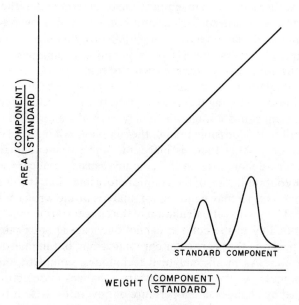

Fig. 5–9.   Relative calibration curve.

IV. Statistical Treatment of Data

A. Definition of Terms

In quantitative analysis, the terms accuracy, precision, repeatability, and reproducibility are frequently used. Accuracy is defined as the measurement of difference between a trial value and the true value. Precision is an expression of exactness or a measure of how well replicate values agree. Accuracy is difficult without precision but precision does not insure accuracy. However, precision with proper calibration gives accuracy. When a single operator repeats an analysis on the same apparatus, repeatability is the term used to define the precision. If the same analysis is run by different operators on different instruments, possibly in different laboratories, the term reproducibility is used to define the precision.

B. Sources of Error

In all measurements, there are errors. Errors can be divided into two categories: indeterminate and determinate. Indeterminate errors are random errors which cannot be eliminated and are inherent in the analytical technique. If minimized, they give high precision. Determinate errors are those errors whose cause and magnitude can be determined. If the determinate errors can be minimized, high accuracy is achieved. However, they can never be completely eliminated. In high pressure liquid chromatography, some determinate errors are: (1) poor sampling techniques; (2) decomposition on the column; (3) change in detector response; (4) recorder performance; (5) calculation errors; and (6) operator prejudice and carelessness.

Errors in poor sampling techniques can result from poor preparation of the sample or improper storage which may cause decomposition or change in concentration. It is possible that all the sample is not injected completely into the column or that there is inaccurate measurement in the syringe. The sample may decompose on the column, causing inaccurate results, or there may be contamination of the sample before injection. It is even possible that the wrong sample may be taken for analysis or the wrong solution may be prepared. The incorrect calculation of the concentration may cause large errors. Overloading of the column cannot only cause poor peak shape and masking of peaks in that particular chromatogram, but in many chromatograms thereafter. As in all analytical techniques, complete and accurate labeling of samples is absolutely necessary. Changes in detector responses may be caused by changes in temperature or flow rates. Also it is important to realize that different detectors give different responses to the same compound and there are different responses by one detector for different compounds. For example, when a UV detector is used, the absorbance is proportional to the molar extinction coefficient. Therefore, for each compound separate calibrations are required. It is also possible that the recorder performance is defective. Calculation and calibration errors are possible sources of inaccuracies. Calculation errors are especially common when there is peak overlap or a drifting baseline. There may be errors in the integrating operation. Another possible source of determinate errors is operator prejudice; the results may be predetermined in the mind of the operator. As was pointed out in Chapter 3, one of the most important factors in obtaining the best performance from a liquid chromatograph is an alert, careful, well-trained technician or operator who has the responsibility for the overall operation.

CHAPTER 6

APPLICATIONS

Interest in nucleic acids, caused by the discovery of DNA, gave great impetus to the development of high pressure liquid chromatography. Although researchers used a variety of techniques, such as thin layer, gas, column, and paper chromatography, and electrophoresis, no one method proved satisfactory in studying these complex, nonvolatile compounds or their components. Spurred by the need for a rapid, accurate, sensitive method for determining the purine and pyrimidine nucleotide, nucleoside, or base concentration of hydrolysates of DNA or RNA, emphasis was placed on formulating methodology for these compounds by high pressure liquid chromatography (Uziel *et al.*, 1968; Horvath *et al.*, 1967; Horvath and Lipsky, 1969c; Burtis and Gere, 1970). It has been found, however, that this chromatographic technique is not limited to nucleic acid components but is valuable in every area of chemical, biochemical, and medical research.

I. General Applications

High pressure liquid chromatography can be used to advantage in many research fields. It can be used to study metabolic pathways and reproductive mechanisms in basic biochemical research, to follow cell growth and reproduction in physiology, and to investigate regenerative processes and diseased tissues in pathology. This type of chromatography is valuable in pharmacology because it is possible to monitor not only the naturally occurring nucleotides, but also to follow the formation of drug metabolites. It is applicable to the study of the metabolism of drugs, the synergistic effects of drugs, the effect of drugs in different species, and it is especially useful in kinetic chemotherapeutic studies. High pressure liquid chromatography can be an invaluable tool in basic research in biology, chemistry, tissue culture work, bacteriology, virology, medicinal chemistry, and biochemical genetics. In synthetic organic research, it is becoming an important analytical tool to follow preparative steps and to analyze for the purity of the final product. In the chemical industry and allied fields, it is important in quality control. It can be used to great advantage in kinetic studies, for example, in studying the kinetics of oxidative phosphorylation. In fact, high speed liquid chromatography can be used to separate, characterize, and identify reactants or products of almost any chemical reaction and is especially valuable in reactions in which the compounds involved are polar, non-volatile, or thermally labile.

Although high pressure liquid chromatography is not yet used routinely in clinical chemistry, toxicology, and medicine, it is predicted that it will replace gas chromatographs for the analysis of all compounds of clinical interest, except those that are gases. Because of its versatility, sensitivity, ease of operation, and simplicity, liquid chromatography can handle most of the compounds of clinical significance, such as steroids, vitamins, fatty acids and their derivatives, barbiturates and other drugs, and their metabolites. Most well-equipped clinical laboratories today possess amino acid analyzers which are specialized high pressure liquid chromatographs with a colormetric detector utilizing a ninhydrin reaction. It may be possible to use liquid chromatographic systems to spot metabolic or genetic defects. It can be used to study drug metabolism in several ways: monitoring the excretion of unreacted drugs, by following the drug metabolite, or by observing the effect of the drug on the naturally occurring nucleotide pools. As a diagnostic tool, high pressure liquid chromatography may be used in diagnosing metabolic disorders or developing tests for predicting these problems. For example, Clifford *et al.* (unpublished data) developed a tolerance test for gout and it was found that AMP induced a characteristic effect in gout-diagnosed patients compared with normal subjects. In nutritional studies, the effect of diet on nucleotide pools may be monitored and in nutritional pharmacology studies, the effect of diet on the metabolism of drugs, can be closely followed.

In the field of ecology, it is evident that high speed liquid chromatography will play an increasingly important role because of its versatility. By this method, minute quantities of materials, such as insecticides, herbicides, antioxidants, and plasticizers, can be assayed. Since it is possible to measure almost any compound rapidly and sensitively by optimizing the operating conditions, the environmental scientist can monitor many impurities in foods, food preservatives, flavors, and pharmaceuticals. It is especially useful in pollution studies and in monitoring pollutants both in air and in water. This is of utmost importance since it has been found that air pollution is a major factor in diseases such as lung cancer, bronchitis, emphysema, and asthma.

In the food industry and in nutritional and food sciences, high speed liquid chromatography can be used to study the adulteration, contamination, and decomposition of foods. It is equally valuable in the pharmaceutical and drug industries for quality control and in the analysis of new products. In toxicology, high pressure liquid chromatography can be used to analyze for minute quantities of barbiturates, narcotics, and other drugs or metals such as bismuth, arsenic, and lead. Although gas chromatography has been used extensively for the analysis of many compounds, it has been found that many of these analyses can be performed better by liquid chromatography. For example, in the analysis of waxes and crude oils, less volatile phenolic mixtures, such as flavenoids and flavones, and in steroids, liquid chromatography is becoming the method of choice. In the analysis of carbohydrates by gas chromatography, the analysis is limited to the methylated saccharides, but liquid chromatography can be used for almost all carbohydrates.

In the fatty acid field, liquid chromatography can handle not only the fatty acids, but larger molecules, such as lipids, phospholipids, sphingolipids, and sulfate lipids. Essential oils, terpines, and terpine derivatives can all be analyzed by high pressure liquid chromatography and proteins and peptides are one of the largest groups that were first analyzed by this technique. High pressure liquid chromatography can handle the separation of alkaloids and vitamins, especially the water-soluble ones, better than gas chromatography and it lends itself more generally to the analysis of antibiotics such as penicillin, the streptomycins, and the polypeptide antibiotics. Furthermore, certain plant pigments such as porphryns and chlorophyll, which are difficult or impossible to separate by gas chromatography, can usually be analyzed by liquid chromatography with the right operating conditions.

II. Specific Applications

A. Nucleic Acid Components

Nucleic acids, when hydrolyzed, break down into the following components: (1) bases (the purine and pyrimidine bases) (Tables 6–1 and 6–2);

TABLE 6–1

PURINES

Purines

Adenine

Guanine

Hypoxanthine

Xanthine

Uric acid

Caffeine

Theobromine

6. APPLICATIONS

TABLE 6–2

PYRIMIDINES

Pyrimidines

Cytosine   Thymine

NH₂   O

in DNA

Cytosine   Uracil

NH₂   O

in RNA

Orotic acid

Intermediate
in pyrimidine
biosynthesis

(2) nucleosides (the *N*-sugar glycosides of these bases (Table 6–3); and (3) nucleotides (the phosphate esters of the nucleosides), oligomers, dimers, and higher nucleotides (Table 6–3). To obtain hydrolysates of RNA for RNA structural and sequence work it is possible to work at either the nucleoside or base level. Uziel *et al.* (1968) chose to work at the nucleoside level "to avoid both the isomeric peaks that arise from acid hydrolysis of nucleic acid materials and the more drastic treatment required to realize bases." Degradation to the nucleoside level is possible either by sodium hydroxide or by enzymic degradation using snake venom diesterase and alkaline phosphatase (Uziel *et al.*, 1968). In work with tRNA they modified the procedure by using sequentially tRNase T₁ and a combination of venom diesterase and alkaline phosphatase to hydrolyze totally tRNA (Uziel and Koh, 1971).

TABLE 6–3

NUCLEOTIDES

	Nucleoside ≡ base + sugar	Nucleotide ≡ base + sugar + phosphate

Nucleotide
≡ base + sugar
+ phosphate
in RNA   DNA
X = OH   H

Bases	Nucleosides	Nucleotides
Adenine	Adenosine	AMP, ADP, ATP
Guanine	Guanosine	GMP, GDP, GTP
Cytosine	Cytidine	CMP, CDP, CTP
Uracil	Uridine	UMP, UDP, UTP
Thymine	Thymidine	TMP, TDP, TTP

To obtain the purine bases from DNA and RNA mild acid hydrolysis has often been used (Loring, 1955). Pyrimidine ribonucleotides are remarkably stable to this mild acid hydrolysis and the hydrolysis of pyrimidine components to the free bases is more difficult. However, more drastic acid hydrolysis conditions, such as the heating of the DNA or RNA with concentrated formic acid at 175°C, with 6 N HCl at 120°C or with 12 N per-

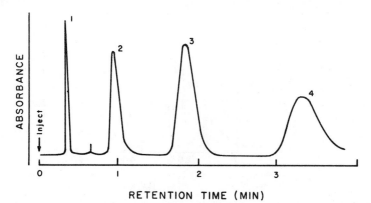

RETENTION TIME (MIN)

Fig. 6–1. Purine and pyrimidine bases. *Instrument*: Varian LCS 1000. *Column*: 300.7 cm long, 1 mm i.d.; packed with pellicular cation exchange resin. *Eluent*: 0.02 M NH$_4$H$_2$PO$_4$, pH 5.5. *Temperature*: 68°C. *Flow rate*: 33.4 ml/hr. *Inlet pressure*: 162 atm. *Sample size*, 800 pmoles of each component: (1) uracil; (2) adenine; (3) guanine; (4) cytosine. *Attenuation*: 0.02 AU full scale. [Contributed by Horvath and Lipsky (1969c).]

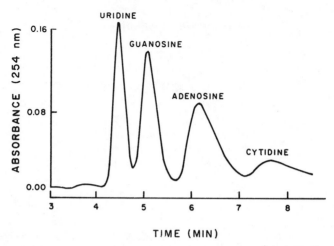

Fig. 6–2. Ribonucleosides. *Instrument*: Varian Aerograph. *Column* 151.7 cm long, 1 mm i.d.; packed with pellicular cation exchange resin. *Eluent*: 0.02 M $NH_4H_2PO_4$, pH 5.6. *Temperature*: 39°C. *Flow rate*: 25.5 ml/hr. *Inlet pressure*: 131 atm. *Sample size*: 300 p moles of each component: adenosine, guanosine, uridine, and cytidine. *Attenuation*: 0.04 AU full scale. [Contributed by Horvath and Lipsky (1969c).]

chloric acid leads to relatively quantitative formation of the free pyrimidine bases. The free components are also present in cells. In order to identify these compounds, chromatograms have been obtained of standard solutions of the purine and pyrimidine bases, their nucleosides and nucleotides. For example, chromatograms of standard solutions of the bases were obtained by Horvath and Lipsky (1969c) (Fig. 6–1). Burtis and Gere (1970) obtained good chromatograms of the ribonucleosides (Fig. 6–2), the 2′,3′-ribonucleotides (Fig. 6–3), the 5′-ribonucleotides (Fig. 6–4) and

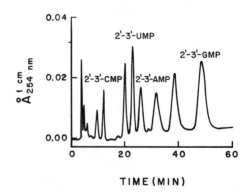

Fig. 6–3. 2′- and 3′-Ribonucleotides. *Instrument*: Varian LCS 1000. *Column*: 193 cm long, 1 mm i.d.; packed with strong, basic pellicular anion exchanger. *Eluent*: linear gradient of KH_2PO_4 from 0.05 to 0.35 M. *Temperature*: 60°C. *Flow rate*: 12 ml/hr. *Inlet pressure*: 51 atm. *Sample*: 1 μg of each isomer mixture (2′-CMP, 3′-CMP, 2′-GMP, 3′-GMP, 2′-AMP, 3′-AMP, 2′-UMP, and 3′-UMP). [Contributed by Horvath *et al.* (1967).]

134

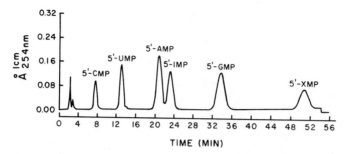

Fig. 6–4. 5'-Ribonucleotides. *Instrument*: Varian LCS 1000, UV detector 0.32. *Separation mode*: pellicular anion exchange with gradient elution. *Column*: 300 cm × 1 mm; packed with pellicular anion exchanger. *Temperature*: 70°C. *Eluent*: 0.01 M KH$_2$PO$_4$ (pH 3.25) to 1.0 M KH$_2$PO$_4$ (pH 4.3). *Flow rate*: 30 ml/hr (column); 15 ml/hr (gradient). *Pressure*: 2750 psi. [Contributed by Burtis and Gere (1970).]

5'-deoxynucleotides (Fig. 6–5). Chromatograms of the mono-, di-, and tri-phosphate nucleotides were obtained by Horvath *et al.* (1967), Burtis and Gere (1970), and by Brown (1970). The standards used by Brown are shown in Fig. 6–6. Burtis and Gere also showed that it is possible to obtain good separation of dinucleotides (Fig. 6–7). Chromatograms have also been obtained of standard solutions of other compounds that may be present

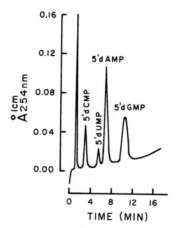

Fig. 6–5. 5'-Deoxynucleotides. *Instrument*: Varian LCS 1000. *Column*: 1 mm × 300 cm at 75°C; packed with pellicular anion exchange resin. *Inlet pressure*: 2800 psi. *Flow rate*: column, 48 ml/hr; gradient, 24 ml/hr. *Starting eluent*: 0.01 M KH$_2$PO$_4$, pH 3.25 *Gradient eluent*: 1.0 M KH$_2$PO$_4$, pH 4.25. *Gradient delay*: none. *Initial volume*: 20 ml. *Sample*: 5'-deoxynucleotides, 1.0 μg each component. *Attenuation*: 0.16 AU full scale. [Contribution by Burtis and Gere (1970).]

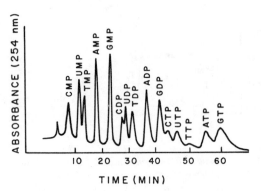

Fig. 6–6. Mono-, di-, and triphosphate nucleotides. *Instrument*: Varian LCS 1000. *Column*: 1 mm × 300 cm; packed with pellicular anion exchange resin. *Flow rates*: 12 ml/hr; 6 ml/hr. *Eluents*: 0.15 M KH$_2$PO$_4$, pH 4.5; 0.25 M KH$_2$PO$_4$ in 2.2 M KCl, pH 4.5. *Initial volume*: 50 ml. *Attenuation*: 0.08 AU full scale. [Contributed by Brown (1970).]

in cells such as the inosine nucleotides, the xanthosine nucleotides, NAD, FMN, FAD, and NADP. The compounds and their retention times, as obtained under the operating conditions used by Brown (1970), are listed in Table 4–1. The workers at Varian have obtained good separation of nucleosides with substituted adenine bases (Fig. 6–8) and also nucleosides with

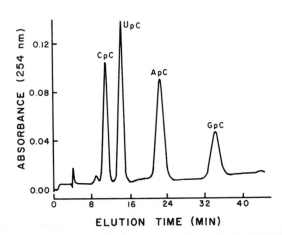

Fig. 6–7. Dinucleotides. *Instrument*: Varian LCS 1000. *Sample*: dinucleotides, 2 μg each. *Inlet pressure*: 1350 psi. *Flow rate*: column, 18 ml/hr; gradient, 9 ml/hr. *Column*: 1 mm × 300 cm at 80°C; packed with pellicular anion exchange resin. *Starting eluent*: 0.01 M KH$_2$PO$_4$, pH 3.30. *Gradient eluent*: 1.0 M KH$_2$PO$_4$, pH 4.2. *Gradient delay*: 7.5 min. *Initial volume*: 50 ml. *Attenuation*: 0.16 AU full scale. [Contribution by Burtis and Gere (1970).]

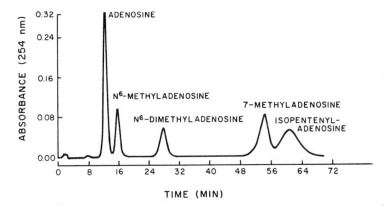

Fig. 6–8. Substituted adenosine nucleosides. *Instrument*: Varian LCS 1000. *Attenuation*: 0.32 AU full scale. *Separation mode*: cation exchange. *Column*: 15 cm × 2.4 mm; packed with Bio-Rad A-7. *Eluent*: 0.4 *M* ammonium formate, pH 4.50. *Temperature*; 60°C. *Flow rate*: 11.4 ml/hr. *Pressure*: 2175 psi. [From Burtis and Gere (1970).]

unusual bases (Fig. 6–9). It has been found that the 3′,5′-cAMP, an important intermediate in hormone action, can be easily separated from 5′-AMP by Brown (1970) (Fig. 6–10). Brooker used different operating conditions to

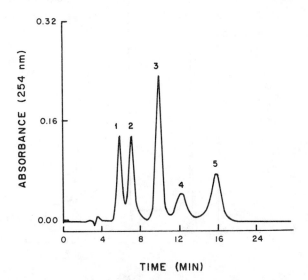

Fig. 6–9. Unusual nucleosides. (1) Pseudouridine; (2) uridine; (3) uracil; (4) xanthosine; (5) 7-methylxanthosine. *Instrument*: Varian LCS 1000, UV detector. *Attenuation*: 0.32 AU. *Separation mode*: cation exchange. *Column*: 15 cm × 2.4 mm; packed with Bio-Rad A-7. *Temperature*: 27°C. *Eluent*: 0.1 *M* ammonium formate, pH 4.50. *Flow*: 7.4 ml/hr. *Pressure*: 2050 psi. [From Burtis and Gere (1970).]

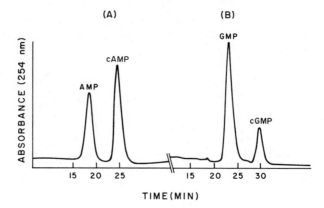

Fig. 6–10. Cyclic AMP and cyclic GMP. *Instrument:* Varian LCS 1000. *Column:* 300 cm × 1 mm; packed with pellicular anion resin. *Starting volume:* 50 ml. *Flow rates:* 12 and 6 ml/hr. *Eluents:* 0.015 *M* KH₂PO₄ and 0.25 *M* KH₂PO₄ in 2.2 *M* KCl. *Samples:* (A) 4 μl AMP and 1 μlc AMP; (B) 4 μl GMP and 1 μlc GMP. *UV output:* 0.16 AU. [Contribution by Brown (1970).]

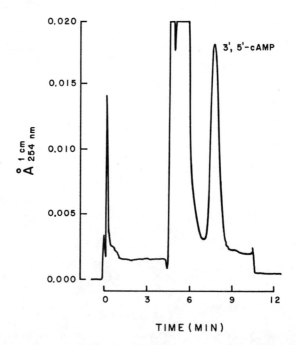

Fig. 6–11. 3′,5′-Cyclic AMP. *Instrument:* Varian LCS 1000; UV detector. *Column:* 300 cm × 1 mm; packed with pellicular anion resin. *Attenuation:* 0.02 AU. *Separation mode:* ion exchange. *Temperature:* 80°C. *Eluent:* 0.01 *M* HCl, pH 2.1. *Flow:* 10 ml/hr. *Pressure:* 400 psi. *Sample:* 30 p moles from a natural sample. [Contribution by Brooker (1970).]

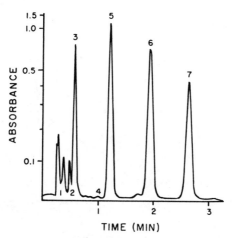

Fig. 6–12. Nucleosides from hydrolysate of mixed *E. coli* B tRNA. *Instrument*: Non-commercial. *Column*: 13 × 0.5 cm; packed with Bio-Rad A-6. *Solvent composition*: 0.8 *M* sodium acetate and 20% ethanol. *Temperature*: 49°C. *Flow rate*: 0.29 ml/min. *Detector*: UV detector (260 nm). *Sample*: 3 nmoles hydrolyzed tRNA. *Peaks*: (1) RNase; (2) pseudouridine; (3) uridine; (4) 4-thiouridine; (5) guanosine; (6) adenosine; (7) cytidine. [Contributed by Uziel and Koh (1971).]

analyze routinely for cAMP in tissues (Fig. 6–11). This assay is sensitive, quantitatively reproducible, and very rapid. The acid-soluble fraction from a hydrolysate of RNA from a cell extract of *E. coli* obtained by Uziel and Koh (1971) is shown in Fig. 6–12.

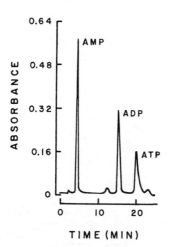

Fig. 6–13. AMP, ADP, and ATP from mitochondria of rat liver. *Instrument*: Varian LCS 1000. *Column*: 2500 × 1 mm; packed with pellicular anion exchanger. *Temperature*: 70°C. *Eluent*: 0.01 *M* KH$_2$PO$_4$ and 0.25 *M* KH$_2$PO$_4$ in 2.2 *M* KCl. *Flow rates*: 24 ml/hr; 12 ml/hr. *Pressure*: 1200 psi. (Contributed by Shmukler (1970b).)

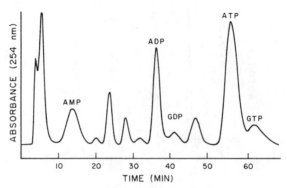

Fig. 6–14. Nucleotides in rat liver extract. *Instrument*: Varian LCS 1000. *Column*: 3 m long, 1 mm i.d.; packed with pellicular anion exchanger (LFS from Varian Aerograph). *Temperature*: 75°C. *Eluents*: 0.015 M KH$_2$PO$_4$, pH 4.5; 0.25 M KH$_2$PO$_4$ in 2.2 M KCl, pH 4.5. *Flow rates*: 12 ml/hr; 6 ml/hr. [Contributed by A. J. Clifford and P. R. Brown (unpublished data).]

Examples of free nucleotides in normal tissues are shown in Fig. 6–13 through 6–23. By optimizing the operating conditions, Shmukler was able to assay routinely and quickly for the adenosine nucleotides in the mitochondria of rat liver (Fig. 6–13) and Clifford *et al.* (1972) obtained highly reproducible chromatograms of nucleotide extracts from homogenized rat

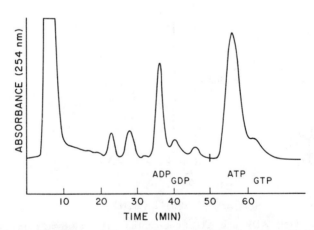

Fig. 6–15. Nucleotide extracts from chicken whole blood. *Instrument*: Varian LCS 1000. *Column*: 3 m long, 1 mm i.d.; packed with pellicular anion exchanger (LFS from Varian Aerograph). *Temperature*: 75°C. *Eluents*: 0.015 M KH$_2$PO$_4$; 0.25 M KH$_2$PO$_4$ in 2.2 M KCl, pH 4.5. *Flow rates*: 12 ml/hr; 6 ml/hr. [Contributed by Brown *et al.* (1972).]

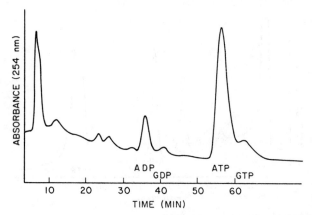

Fig. 6–16. Nucleotide extracts from rabbit whole blood. *Instrument*: Varian Aerograph. *Column*: 3m long, 1 mm i.d.; packed with pellicular anion exchanger (LFS from Varian Aerograph) *Temperature*: 75°C. *Eluents*: 0.015 M KH$_2$PO$_4$; 0.25 M KH$_2$PO$_4$ in 2.2 M KCl, pH 4.5. *Flow rates*: 12 ml/hr; 6 ml/hr. [Contributed by Brown *et al.* (1972).]

liver (Fig. 6–14). The nucleotide profiles of the whole blood of man and many animals were obtained by Brown *et al.* (1972). This information is important in enzyme studies, in investigating metabolic pathways, and in deciding which animal is most suitable for use in biochemical, pharmacological, and medical studies. Illustrations of the nucleotide patterns in the whole blood of chicken, rabbit, cow, and horse are shown in Fig. 6–15 through 6–18. Because of the great sensitivity of the high pressure liquid

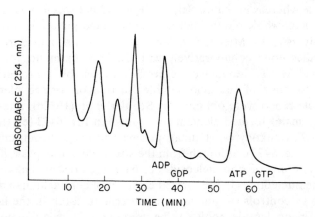

Fig. 6–17. Nucleotide extracts from cow whole blood. *Instrument*: Varian LCS 1000. *Column*: 3m long, 1 mm i.d.; packed with pellicular anion exchanger (LFS from Varian Aerograph). *Temperature*: 75°C. *Eluents*: 0.015 M KH$_2$PO$_4$; 0.25 M KH$_2$PO$_4$ in 2.2 M KCl, pH 4.5 [Contributed by Brown *et al.* (1972).]

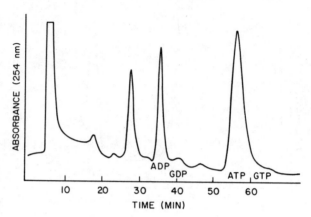

Fig. 6–18. Nucleotide extracts from horse whole blood (See Fig. Legend 6–17 for an explanation).

chromatographic system, it was possible to detect small differences in the patterns. Although many of the chromatograms are similar, each species has a characteristic profile. Within a species, however, no significant differences could be observed.

The reproducibility of chromatograms within a species is demonstrated in Fig. 6–19 and Fig. 6–20. In Fig. 6–19, Clifford's chromatogram of human whole blood, the UV output was attenuated for the ATP peak so that it would stay on scale whereas in that of Scholar (Fig. 6–20) it was not. In numerous other studies on whole blood from normal volunteers in our laboratories at Brown University by Miech, Scholar, Nelson, Brown, and Parks and in several studies done collaboratively with Clifford, Riumallo, Munro, and Scrimshaw of M.I.T. the nucleotide profiles are reliably reproducible.

Chromatograms of the acid soluble fraction of formed elements in human whole blood were obtained by Scholar et al. (1972a). Examples of chromatograms of formed elements are shown in Fig. 6–21 (erythrocytes) and Fig. 6–22 (leukocytes). Human plasma had insignificant amounts of nucleotides (Fig. 6–23), although from the size of the initial peak, it appears that large amounts of ribonucleosides or bases are present. The chromatograms of the nucleotides in the formed elements of normal human blood are being used as controls or standards in current research in the hope that differences in nucleotide pools can be used clinically as a diagnostic tool to spot potential or actual metabolic or neoplastic problems.

In a study to devise a diagnostic test for persons in a population with a tendency for gout, Clifford et al. (1971) investigated the nucleotide pools of whole blood of normal volunteers, hyperuracemic, and gout-diagnosed patients. The nucleotide pools in blood of a gout patient is shown in Fig. 6–24. There were no differences in the qualitative nucleotide patterns of normal and gout patterns, but quantitatively the ATP of the normal

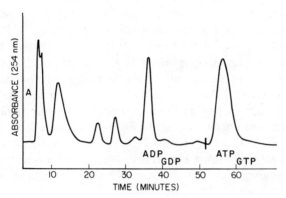

Fig. 6–19. Nucleotide extracts from human whole blood. *Instrument*: Varian LCS 1000. *Column*: 3 m long, 1 mm i.d.; packed with pellicular anion exchanger (LFS from Varian Aerograph). *Temperature*: 75°C. *Eluents*: 0.015 M KH$_2$PO$_4$, pH 4.5; 0.25 M KH$_2$PO$_4$ in 2.2 mKCl, pH 4.5. *Flow rates*: 12 ml/hr, 6 ml/hr. *Extraction method*: perchloracetic acid (Contributed by Clifford *et al.*, 1972).

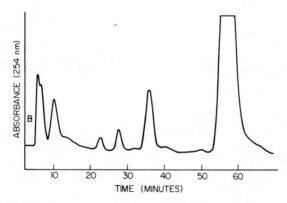

Fig. 6–20. Nucleotide extracts from human whole blood. *Instrument*: Varian LCS 1000. *Column*: 3 m long, 1 mm i.d.; packed with pellicular anion ex changer (LFS from Varian Aerograph). *Temperature*: 75°C. *Eluents*: 0.015 M KH$_2$PO$_4$, pH 4.5; 0.25 M KH$_2$PO$_4$ in 2.2 M KCl, pH 4.5. *Flow rates*: 12 ml/hr, 6 ml/hr. *Extraction method*: trichloracetic acid (Contributed by Scholar *et al.*, 1972a).

volunteers seemed to be lower than that of the gout patients. The comparison of UV-absorbing constituents in normal urine and that from a patient with the Lesch-Nyan syndrome is shown in Fig. 6–25.

Gell, in investigating the changes that occur during the reproductive process of sea urchins, found that the nucleotide patterns of sea urchin sperm and ova differed markedly (Figs. 6–26 and 6–27). Senft and co-workers (1972), in their studies on schistosomiasis, investigated the nucleotide pools of *Schistosoma mansoni* (Fig. 6–28) and the chromatogram of a nucleotide extract of murine leukemia cells L5185Y prepared by Chu of the Roger Williams Hospital and Brown University is shown in Fig. 6–29.

In pharmacological studies, the effect of a drug on the naturally occurring nucleotide pools as well as the formation of the analog nucleotide can be followed. Scholar, *et al.* (1972) studied *in vitro* the synergistic effect of 6-mercaptopurine and 6-methylmercaptopurine riboside on Sarcoma 180 cells. The chromatogram in Fig. 6–30 is of the Sarcoma 180 cells, and that in Fig. 6–31 is of Sarcoma 180 cells after treatment by the drugs. Nelson and Parks of Brown University (1972) investigated the synergistic effect of thioguanine and 6-methylmercaptopurine riboside on Sarcoma 180

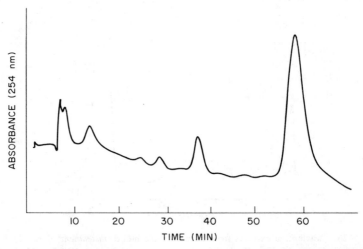

Fig. 6–21. Nucleotide extracts from human erythrocytes. *Instrument*: Varian LCS 1000. *Column*: 3 m long, 1 mm i.d.; packed with pellicular anion exchanger (LFS from Varian Aerograph). *Temperature*: 75°C. *Eluents*: 0.015 M KH$_2$PO$_4$; 0.25 M KH$_2$PO$_4$ in 2.2 M KCl, pH 4.5. *Flow rates*: 12 ml/hr; 6 ml/hr. [Contributed by Brown, unpublished data.]

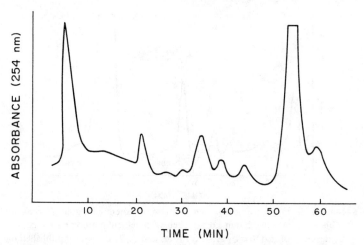

Fig. 6–22. Nucleotide extracts from human leukocytes. *Instrument*: Varian LCS 1000. *Column*: 3 m long, 1 mm i.d.; packed with pellicular anion exchanger (LFS from Varian Aerograph). *Temperature*: 75°C. *Eluents*: 0.015 M KH$_2$PO$_4$; 0.25 M KH$_2$PO$_4$ in 2.2 M KCl, pH 4.5. *Flow rate*: 12 ml/hr.; 6 ml/hr. [Contributed by Scholar *et al.* (1972a).]

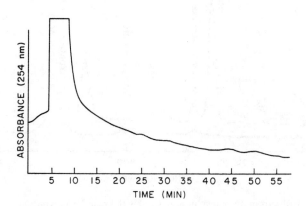

Fig. 6–23. Nucleotide extracts from human plasma. *Instrument*: Varian LCS 1000. *Column*: 3 m long, 1 mm i.d.; packed with pellicular anion exchanger (LFS from Varian Aerograph). *Temperature*: 75°C. *Eluents*: 0.015 M KH$_2$PO$_4$; 0.25 M KH$_2$PO$_4$ in 2.2 M KCl, pH 4.5. *Flow rate*: 12 ml/hr; 6 ml/hr. [Contributed by Scholar *et al.* (1972a).]

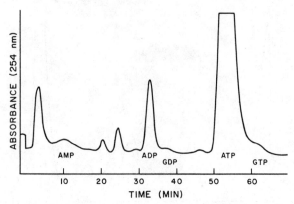

Fig. 6–24. Nucleotide extracts from whole blood of patient with gout. *Instrument*: Varian LCS 1000. *Column*: 3 m long, 1 mm i.d.; packed with pellicular anion exchanger (LFS from Varian Aerograph). [Contributed by A. J. Clifford, and P. R. Brown (unpublished data).]

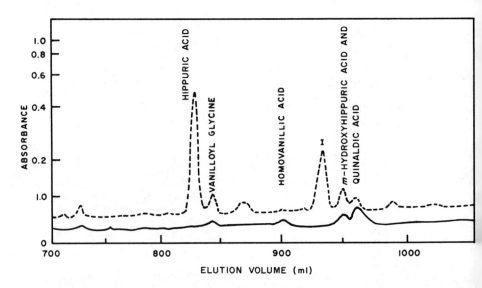

Fig. 6–25. Urine from normal subject and one with Lesch-Nyan syndrome. (– –) Normal, 260 nm (—) Lesch-Nyan, 260 nm. *Instrument*: Varian LCS 1000. *Column*: 0.24 × 100 cm; packed with Aminex A-27 strong anion exchanger. *Eluent*: 0.015 M sodium acetate, pH 4.40; 6.0 M sodium acetate, pH 4.4. *Flow rate*: 8 ml/hr; 4 ml/hr. *Temperature*: 60°C. [Contributed by Jolley and Scott (1970a).]

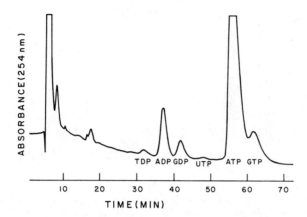

Fig. 6–26. Nucleotide extracts from sea urchin sperm. *Instrument*: Varian LCS 1000. *Column*: 3 m long, 1 mm i.d.; packed with pellicular anion exchanger (LFS from Varian Aerograph). *Temperature*: 75°C. *Eluents*: 0.015 M KH_2PO_4; 0.25 M KH_2PO_4 in 2.2 M KCl. pH 4.5. *Flow rates*: 12 ml/hr; 6 ml/hr. [Contributed by J. Gell and Brown (unpublished data).]

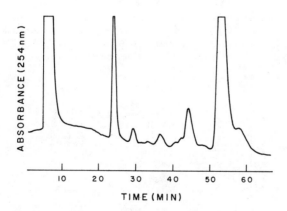

Fig. 6–27. Nucleotide extracts from sea urchin ova. *Instrument*: Varian LCS 1000. *Column*: 3 m long, 1 mm i.d.; packed with pellicular anion exchanger (LFS from Varian Aerograph). *Temperature*: 75°C. *Eluents*: 0.015 M KH_2PO_4; 0.25 M KH_2PO_4 in 2.2 M KCl. *Flow rate*: 12 ml/hr.; 6 ml/hr. [Contributed by Gell and Brown (unpublished data).]

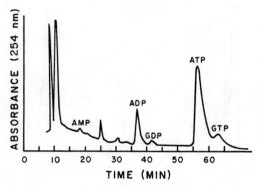

Fig. 6–28. Nucleotide extracts from *Schistosoma mansoni*. *Instrument*: Varian LCS 1000. *Column*: 3 m long, 1 mm i.d.; packed with pellicular anion exchanger (LFS from Varian Aerograph). *Temperature*: 75°C. *Eluents*: 0.015 M KH_2PO_4; 0.25 M KH_2PO_4 in 2.2 M KCl. *Flow rate*: 12 ml/hr.; 6 ml/hr. [Contributed by Senft *et al*. (1972).]

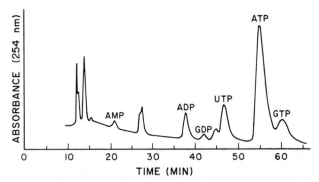

Fig. 6–29. Nucleotide extracts from murine leukemia cells. *Instrument*: Varian LCS 1000. *Column*: 3 m long, 1 mm i.d.; packed with pellicular anion exchanger (LFS from Varian Aerograph). *Temperature*: 75°C. *Eluents*: 0.015 M KH_2PO_4; 0.25 M KH_2PO_4 in 2.2 M KCl. 4.5. *Flow rate*: 12 ml/hr.; 6 ml/hr. [Contributed by Brown (1970).]

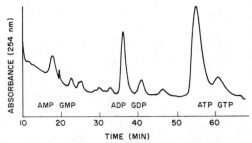

Fig. 6–30. Nucleotide extracts from Sarcoma 180 cells. *Instrument*: Varian LCS 1000. *Column*: 3 m long, 1 mm i.d.; packed with pellicular anion exchange (LFS from Varian Aerograph). *Temperature*: 75°C. *Eluents*: 0.015 M KH_2PO_4; 0.25 M KH_2PO_4 in 2.2 M KCl, pH 4.5. *Flow rate*: 12 ml/hr.; 6 ml/hr. [Contributed by Brown (1970).]

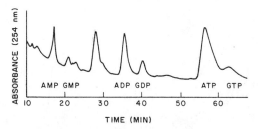

TIME (MIN)

Fig. 6–31. Nucleotide extracts from Sarcoma 180 cells treated with 6-mercaptopurine and 6-methylmercaptopurine riboside. *Instrument*: Varian LCS 1000. *Column*: 3 m long, 1 mm i.d.; packed with pellicular anion exchanger (LFS from Varian Aerograph). *Temperature*: 75°C. *Eluents*: 0.015 M KH$_2$PO$_4$; 0.25 M KH$_2$PO$_4$ in 2.2 M KCL, pH 4.5. *Flow rate*: 12 ml/hr.; 6 ml/hr. [Contributed by Brown (1970).]

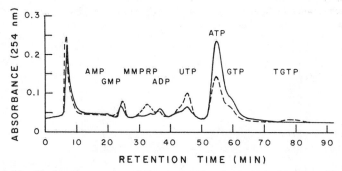

RETENTION TIME (MIN)

Fig. 6–32. Nucleotide extracts of Sarcoma 180 cells treated with thioguanine and 6-methylmercaptopurine riboside. (—) No MMPR; (– –) MMPR. *Instrument*: Varian LCS 1000. *Column*: 3 m long, 1 mm i.d.; packed with pellicular anion exchanger (LFS from Varian Aerograph). *Temperature*: 75°C. *Eluents*: 0.015 M KH$_2$PO$_4$; 0.25 M KH$_2$PO$_4$ in 2.2 M KCl, pH 4.5. [Contributed by Nelson and Parks (1972).]

cells and the cells after treatment are shown in Fig. 6–32. The peaks not previously seen in the chromatogram of the Sarcoma 180 cells are the phosphorylated analogs.

Human erythrocytes were used as a model system to study the metabolic pathways of some of the purine antimetabolites (Preliminary Report, Brown and Parks, 1971) in an attempt to find the mechanism of their action as antitumor agents. Although 6-methylmercaptopurine riboside and 6-mercaptopurine formed the monophosphate when incubated with fresh red blood cells, fluoradenine, fluoroadenosine, and tubercidin readily became phosphorylated up to the triphosphate levels when incubated under the same conditions (Figs. 6–33). Moreover, the concentration of the naturally occurring adenosine and guanosine nucleotides did not change. With schistosomes, on incubation with fluoroadenosine or tubercidin (Fig. 6–34)

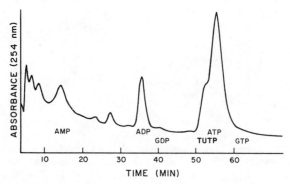

Fig. 6–33. Nucleotide extracts from human erythrocytes incubated with tubercidin sine. *Instrument*: Varian LCS 1000. *Column*: 3 m long, 1 mm i.d.; packed with pellicular anion exchanger (LFS from Varian Aerograph). *Temperature*: 75°C. [Contributed by P. R. Brown and R. E. Parks, Jr. (unpublished data).]

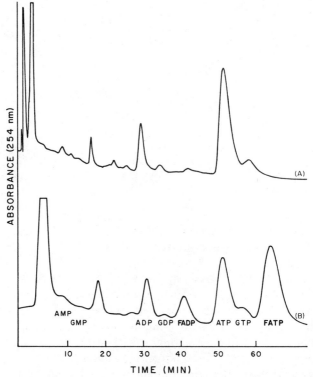

Fig. 6–34. Nucleotides in *Schistosoma mansoni* incubated with 2-fluoroadenosine. *Instrument*: Varian LCS 1000. *Column*: 1 mm × 3 m; packed with pellicular anion exchange resin. *Eluents*: 0.015 M KH_2PO_4, pH 4.5; 0.25 M KH_2PO_4 in 2.2 M KCl, pH 4.5. *Flow rates*: 12 ml/hr, 6 ml/hr. *Temperature*: 75°C. *Starting volume*: 50 ml. [Contributed by Stegman *et al.* (1972).]

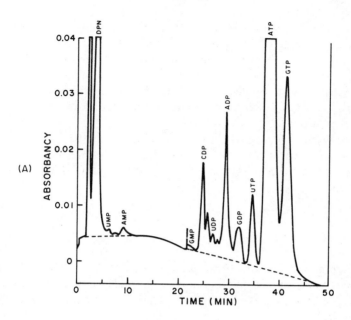

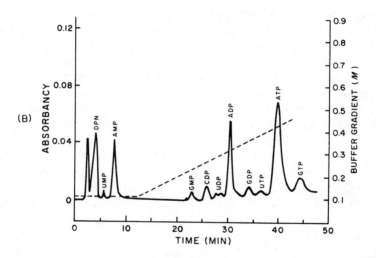

Fig. 6–35. Nucleotides in rat brain. *Instrument*: Varian LCS 1000. *Column*: 1 mm × 3 m; packed with pellicular anion exchange resin. *Eluents*: 0.01 *M* KH₂PO₄ + 0.001 *M* H₃PO₄; 0.25 *M* KH₂PO₄ in 2.2 *M* KCl. *Gradient delay*: 20 minutes. *Starting volume*: 45 ml. *Flow rates*: 24 ml/hr, 12 ml/hr. [Contributed by Shmukler (1972).]

Senft *et al.* (1972) also found that both of these analogs readily form the triphosphate nucleotide. High pressure liquid chromatography has been found to be an invaluable tool in these studies because of its high resolution, sensitivity, and speed. Moreover, it is possible to monitor not only the effect of purine and pyrimidine antimetabolites on the naturally occurring nucleotide pools, but it is also possible to do time studies on the phosphorylation of these drugs.

In a study on fatigue stress in rats Shmukler *et al.* (1972a) were able to obtain excellent chromatograms of the acid-soluble nucleotides in rat brain. A chromatogram at a UV output of 0.32 AU and at 0.04 AU are shown in Fig. 6–35.

B. COMPOUNDS OF BIOCHEMICAL INTEREST

Other compounds of biochemical importance that have been separated by high pressure liquid chromatography are steroids, vitamins, amino acids, lipids, and carbohydrates. Steroids separated by Henry *et al.* (1971b) are shown in Fig. 6–36, by Siggia and Dishman (1970) in Fig. 6–37, and by the Varian group in Fig. 6–38. Chromatograms of three molting insect steroids were obtained by workers at Waters Associates (Fig. 6–39). Vitamins A, D, and E can be readily separated as is shown in Fig. 6–40. Although amino acids are separated routinely by amino acid analyzers, which are specialized liquid chromatographic systems with a ninhydrin reagent for detecting the acids, they can also be separated by high pressure liquid chromatography. Gere of Varian Associates obtained chromatograms of aspartic acid, proline, valine, and phenylalanine using a microadsorption detector (Fig. 6–41). Free fatty acids and cholesterol esters were separated from cholesterol and total glycerides by researchers at Waters Associates (1970) (Fig. 6–42) and the glycerides, tri-, di-, and monopalmitin were well resolved by Woods and Lantz (1970) of Barber-Coleman using a flame ionization detector (Fig. 6–43). Carbohydrates as well as lipids can be detected using a microadsorption detector. This was demonstrated by Gere of Varian Associates (Fig. 6–44).

High pressure liquid chromatography can be a powerful tool in enzymic assays. A method for the assay of 3',5'-cAMP-phosphodiesterase has been described by Pennington (1971) based on Brooker's (1970) separation of AMP and 3',5'-cAMP. It is possible to use the reverse of the enzymic peak-shift technique for enzymic assays. If a known amount of a substrate is added to an enzyme sample, the amount of enzyme present can be calculated from the amount of product formed. For example, if a sample of hexokinase is added to a reaction mixture of an excess of glucose and a known amount of

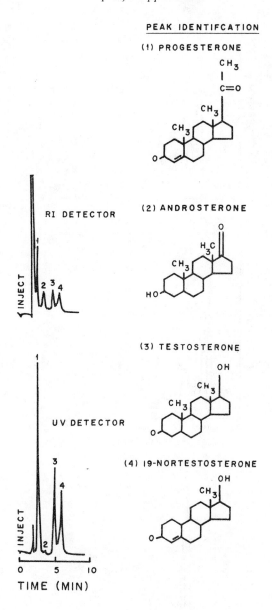

Fig, 6–36. Steroids. *Instrument*: du Pont 820 liquid chromatograph. *Column*: 2.1 mm × 1 m; packed with 1% BOP on Zipax. *Eluent*: heptaine. *Flow rate*: 1 ml/min. *Pressure*: 600 psi. *Temperature*: ambient. *UV output*: 0.32 AUFS, RI 8 × 10⁻⁵ RIFS. *Sample*: (1) progesterone; (2) andosterone; (3) testosterone; (4) 19-nortestosterone. [Contributed by Henry *et al.* (1971b).]

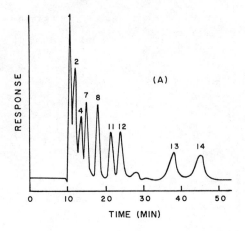

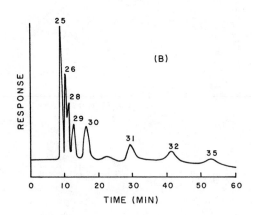

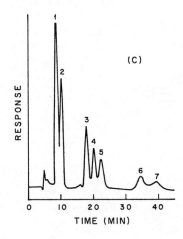

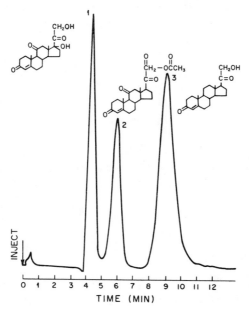

Fig. 6–38. Reverse phase chromatography of steroids. *Instrument*: Varian LCS 4100. *Column*: 75 cm × 2.4 mm; packed with Permaphase ODS. *Solvent*: 20% MeOH/80% H_2O. *Flow rate*: 30 ml/hr. *Pressure*: 120 psi. *Sample*: (1) 0.3 μg cortisone; (2) 0.2 μg cortisone acetate; (3) 0.3 μg Deoxycorticosterone. [Contributed by Varian Associates (1971).]

◀ **Fig. 6–37.** Steroids. (A) *Instrument*: noncommercial. *Column*: 485 mm × 2 mm; packed with 28% LA 1 on CTFE. *Separation mode*: liquid-liquid partition. *Temperature*: ambient. *Eluent*: water. *Flow rates*: programmed initially 0.1 ml/min increased to 0.13, 0.21, 0.26, 0.44 ml/min at A, B, C, D. *Detector*: UV. (1) 6$_β$-hydroxycortisone; (2) aldosterone; (3) Reichstein's U; (4) cortisone; (5) Reichstein's E; (6) cortisol; (7) 11-dehydro-corticosterone; (8) corticosterone; (9) 4-pregnene-11$_B$, 17-diol-3, 20 dione; (10) 4-pregnene-17a-ol-3, 11, 20-trione; (11) 11-deoxycortisol; (12) cortisone 21-acetate; (13) 4-pregnene-20$_B$, 21-diol-3-one; (14) deoxycorticosterone.

(B) *Instrument*: noncommercial. *Column*: 485 mm × 2 mm; packed with 28% LA 1 on CTFE. *Separation mode*: liquid-liquid partition. *Temperature*: ambinet. *Eluent*: water adjusted to pH 11.5 with NaOH. *Flow rates*: programmed initially 0.1 ml/min increased to 0.13, 0.21, 0.26, 0.44 ml/min at A, B, C, D. *Detector*: UV. (25) estradiol-17-glucosiduronic acid; (26) estriol; (27) 17-epiestriol; (28) 16-ketoestrone; (29) 16-ketoestradiol; (30) 16-ketoestradiol; (31) equilinen; (32) estradiol; (33) 2-methoxyestrone; (34) equilin; (35) 17-estrone.

(C) *Instrument*: noncommercial. *Column*: 485 mm × 2 mm; packed with 28% LA 1 on CTFE. *Separation mode*: liquid-liquid partition. *Temperature*: ambient. *Eluent*: water, *Flow*: initial 0.17 ml/min, increased to 0.49 ml/min at A. *Detector*: UV. (1) 4-Androstene-3, 11, 17-trione; (2) 4-androstene-11 B-ol-3, 17-dione: (3) 1,4-androstadiene-17$_B$-ol-3-one; (4) 19-nor-4-androstene-2, 17-dione; (5) 19-nor-testosterone; (6) 4-androstene-3, 17-dione; (7) testosterone. [Contributed by Siggia and Dishman (1970).]

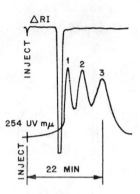

Fig. 6–39. Steroids. *Instrument*: Waters ALC-100. *Solvent*: methanol: water (7:3). *Flow rate*: 3.1 ml/min. *Column*: 3ft × 0.305 in. i.d.; packed with Poragel PN. *Sample load*, 50 μg, total: (1) ecdysone; (2) cyasterone; (3) ponasterone. [Contributed by Waters Associates (1970).]

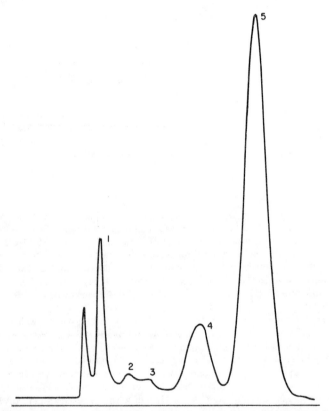

Fig. 6–40. Vitamins A and E. *Instrument*: Varian LCS 1000. *Column*: 50 cm × 2.4 mm; packed with Corasil II. *Eluent*: 1% isopropyl alcohol in isooctane. *Flow rate*: 14 ml/hr. *Sample*: (1) vitamin E; (2) unknown; (3) unknown; (4) retinol isomer (perhaps 13-*cis*); (5) retinol (all-*trans*). [Contributed by Varian Associates (1971).]

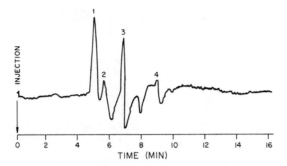

Fig. 6–41. Amino acids. *Instrument*: Varian LCS 1000. *Column*: 30 cm × 0.6 cm; packed with resin UR30. *Eluent*: 0.2 *N* sodium citrate, pH 3.25. *Flow rate*: 25 ml/hr. *Sample*: (1) aspartic acid; (2) proline; (3) valine; (4) phenylalanine. [Contributed by D. Gere of Varian Associates (1971).]

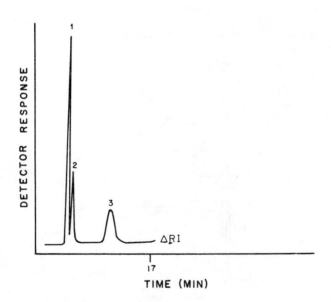

Fig. 6–42. Lipids. *Instrument*: Waters ALC/GPC 301. *Column*: 68 in. × 2.3 mm i.d.; packed with Corasil I. *Solvent*: isooctane:isopropyl ether (90:10). *Flow rate*: 1 ml/min. *Sample*: (1) free fatty acids and cholesterol esters; (2) total glycerides; (3) cholesterol. [Contributed by Waters Associates, Inc. (1970).]

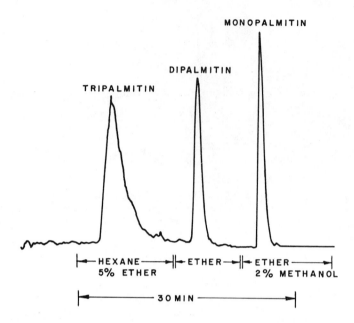

Fig. 6–43. Lipids. *Instrument*: *Column*: 0.35 × 10 cm; packed with Florasil. *Eluent*: stepwise (1) hexane, 5% Et_2O; (2) Et_2O; (3) Et_2O 2% MeOH. *Flow rate*: 0.5 ml/min. *Samples*: tripalmitin, dipalmitin, and monopalmitin. [Contributed by R. A. Woods and C. D. Lantz (1970), Barber-Coleman.]

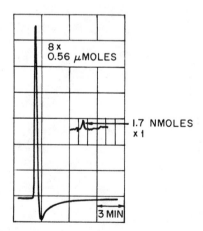

Fig. 6–44. Carbohydrates. *Instrument*: Varian LCS 1000. *Column*: adsorbant. activated charcoal. *Eluent*: water. *Flow rate*: 15 ml/hr. [Contributed by D. Gere of Varian Associates (1970).]

ATP, the activity of the hexokinase can be calculated from either the rate of formation of ADP or decrease in concentration of ATP. Stahl *et al.* (1972) designed a high pressure liquid chromatograph for the analysis of tricarboxylic acid cycle intermediates and related compounds and Stevenson (1971) was able to separate complex organic mixtures by high resolution chromatography.

C. DRUGS AND THEIR METABOLITES

Before the advent of high pressure liquid chromatography, the assay of many drugs or their metabolites required either multiple separation steps or complicated methods. For many of these analyses, liquid chromatography provides a rapid, simple, sensitive, and quantitative technique. For example, Anders and Latorre (1970) separated five barbital compounds using a gradient eluent with an anion exchange column (Fig. 6–45). Waters Associates workers used dual detectors (UV and differential refractometer) to obtain standards of 12 of these compounds, using partition chromatography with chloroform as the eluent. As an example, the chromatogram of barbital is shown in Fig. 6–46. Anders and Latorre obtained good separation of diphenylhydantoin and its metabolites (Fig. 6–47), of the glucuronide and sulfate conjugates of *p*-nitro phenol (Fig. 6–48) and parahydroacetanilide metabolites in human urine (Fig. 6–49). A chromatogram of the anal-

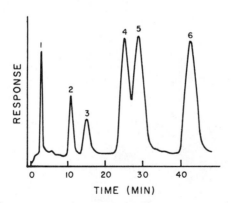

Fig. 6–45. Barbiturates. *Instrument*: Varian LCS 1000. *Separation mode*: anion exchange with linear gradient. *Column*: 300 × 1 mm; packed with pellicular anion resin. *Eluent*: 0.1 to 1.0 *M* NaCL, pH 7. *Temperature*: 80°C. *Flow*: 26 ml/hr. *Pressure*: 700–900 psi. *Sample*: (1) ketohexobarbital (0.3 μg); (2) hydroxyamobarbital (13 μg); (3) contaminant; (4) amobarbital (5.3 μg); (5) phenobarbital (2.5 μg); (6) hydroxyphenobarbital (3.5 μg). [Contributed by Anders and Latorre (1970).]

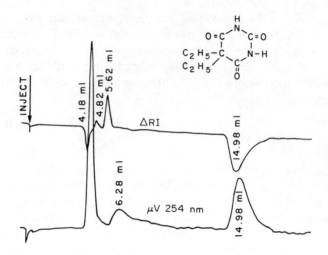

Fig. 6–46. Barbiturates. *Instrument*: Waters ALC 202/401. *Column*: 6 ft × 2.3 mm; packed with Corasil II. *Eluent*: chloroform. *Flow rate*: 0.48 ml/min. [Contributed by Waters Associates, Inc. (1970).]

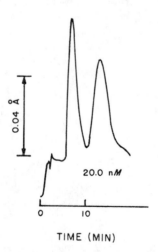

TIME (MIN)

Fig. 6–47. Diphenylhydantoin and metabolites: diphenylhydantoin and hydroxy-diphenylhydatoin. *Instrument*: Varian LCS 1000, UV detector. *Column*: 300 cm × 1 mm; packed with pellicular anion exchange resin. *UV attenuation*: 0.04 AU. *Separation mode*: anion exchange. *Temperature*: 80°C. *Eluent*: 0.02 M KH$_2$PO$_4$, pH 4.5. *Flow rate*: 36.3 ml/hr. [Contributed by Anders and Latorre (1970).]

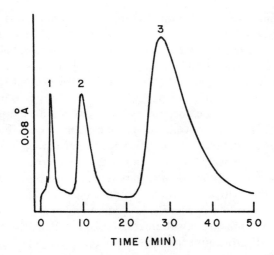

Fig. 6–48. Drug conjugates, glucuronides, and sulfates. *Instrument*: Varian LCS 1000. *Column*: 250 cm × 1 mm; packed with a pellicular anion exchange resin. *Temperature*: 80°C. *Eluent*: 10.0 mM formic acid, pH 3, containing 1.0 M KCL. *Flow rate*: 30 ml/hr. *Pressure*: 800–1000 psig. *Sample*: (1) *p*-nitrophenylglucuronide (50 nmoles); (2) *p*-nitrophenol (100 nmoles); (3) *p*-nitrophenyl sulfate (150 nmoles). [Contributed by Anders and Latorre (1971).]

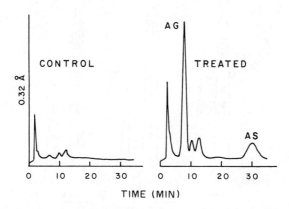

Fig. 6–49. Metabolites in human urine. *Instrument*: Varian LCS 1000. *Column*: 250 cm × 1 mm; packed with a pellicular anion exchange resin. *Temperature*: 80°C. *Eluent*: Gradient 1.0 mM formic acid, pH 4; 1.0 mM formic acid, pH 4, containing 2.0 M KCL. *Sample*: 1 µl of ultrafiltrate from control and from normal male who had taken 15 mg *p*-hydroxyacetanilide (Tylenol) orally. AG = acetanilide glucuronide; AS = acetanilide sulfate. [Contributed by Anders and Latorre (1971).]

gesics, aspirin, N-acetyl-p-aminophenol and caffeine was obtained by workers at Nester/Faust (Fig. 6–50), and good separation of acetylsalicylic acid, caffeine, and phenacetin was achieved by researchers at du Pont (Fig. 6–51). C. G. Scott and Bommer (1970) obtained good resolution of benzodiazepine and its metabolites (Fig. 6–52). The decongestants in cough medicines, phenylephrine hydrochloride, acetaminophen, dextromethorphan hydrobromide, and chlorphiramine maleate were separated by Waters Associates using sodium citrate as an eluent and a weak basic anion exchange resin in the column (Fig. 6–53). The same group separated morphine, heroin, and quinine using both UV and differential refractometer detectors (Fig. 6–54), and also obtained a standard for erythromycin (Fig. 6–55). With the sulfonamides, which are potent, broad-range bacteriostatic agents, it is important to control dosage levels. It is common to use a trisulfapyrimidine combination of sulfadazine, sulfamerazine, and sulfamethazine. Although the current USP procedure for analysis of these drugs is a paper chroma-

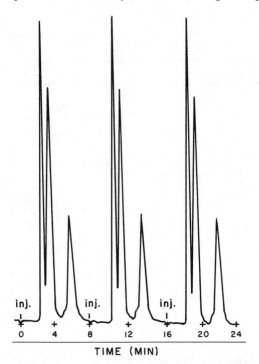

TIME (MIN)

Fig. 6–50A. Analgesics (aspirin, N-acetyl p-aminophenol, caffeine). *Instrument*: Nester/Faust Model 1200 liquid chromatograph. *Column*: 3 mm × 50 cm; packed with Sil-x adsorbent. *Solvent*: 1:1, THF:isopropyl ether. *Flow rate*: 0.82 ml/min. *Pressure*: 150 psi. *Detector*: UV. *Attenuation*: 0.10 o.d. (Contributed by Perkin-Elmer Corp., formerly Nester/Faust Mfg. Inc.)

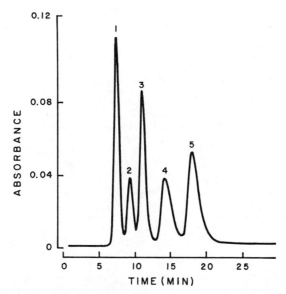

Fig. 6–50B. Analgesics. *Instrument*: Varian Aerograph LCS 1000. *Column*: 300 cm ×
1 mm; pellecular anion. *Eluent*: 1.0 m TRIS; pH 9.0. *Flow rate*: 8.6 ml/hr. *Samples*: (1) caffeine,
(2) asperin (3) 4-hydroxyacetamilide (4) phenacetin (5) solicylamide. [Contributed by C. A.
Burtis (1970a).

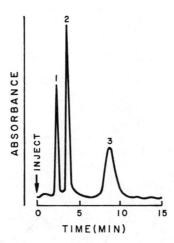

Fig. 6–51. Analgesics. *Instrument*: du Pont 820 liquid chromatograph. *Column*: 1 m ×
2.1 mm; strong cation exchange (SCX). *Mobile phase*: water. *Pressure*: 1200 psig. *Detector*:
UV photometer at 254 nm. *Peak identity*: (1) acetylsalicylic acid; (2) caffeine; (3) phenacetin.
(Contributed by du Pont Instrument Products Division.)

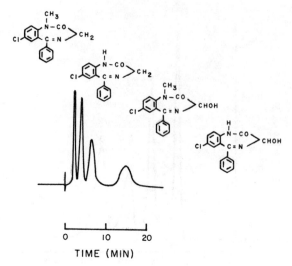

TIME (MIN)

Fig. 6–52. Benzodiazepine and metabolites. *Instrument*: noncommercial, UV detector. *Separation mode*: liquid–liquid partition. *Column*: 100 cm × 1 mm; packed with OPN DuraPak. *Eluent*: 10% isopropanol in hexane. [Contributed by C. G. Scott and Bommer (1970).]

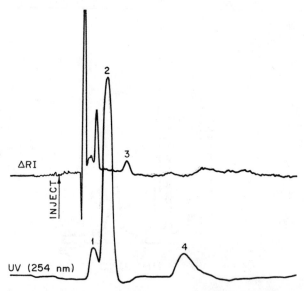

Fig. 6–53. Decongestants. *Instrument*: Waters ALC 202/401. *Solvent*: 0.35 N sodium citrate, pH 5.28. *Flow rate*: 0.50 ml/min. *Temperature*: 30°C. *Column*: 4 ft × 2.3 mm i.d.; weak basic anion secondary amine exchanger. *Components*: (1) phenylephrine—HCl; (2) acetaminophen; (3) dextromethorphan—HBr (4) chlorpheniramine maleate. (Contributed by Waters Associates (1970).)

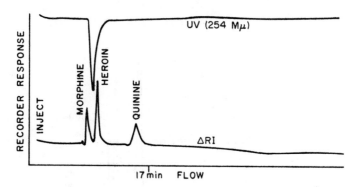

Fig. 6–54. Morphine, heroin, and quinine. *Instrument*: Waters ALC 202/401. *Column*: 6 ft × 2 mm i.d.; packed with Corasil II. *Solvent*: chloroform. *Flow rate*: 0.5 ml/min. *Pressure*: 300 psig. *Sample*: 1 mg. *Detectors*: both UV and RI. [Contributed by Waters Assoc. (1970).]

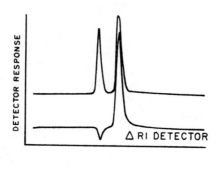

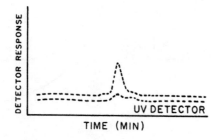

Fig. 6–55. Erythromycin. *Instrument*: Waters ALC-100. *Column*: 3 ft × 2 mm i.d.; packed with Corasil II. *Solvent*: chloroform. *Flow rate*: 0.42 ml/min. [Contributed by Waters Assoc., Inc. (1970).]

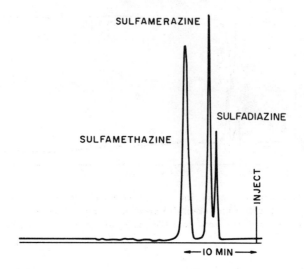

Fig. 6–56. Sulfa drugs (trisulfapyrimidines). *Instrument*: Waters ALC-202. *Column*: 6 ft × 2.3 mm i.d.; packed with WBAX-10. *Solvent*: Water. *Flow rate*: 0.5 ml/min. *Pressure*: 1500 psi. *Sample*: 0.8 μg "SULFA." [Contributed by Waters Associates, Inc. (1970).]

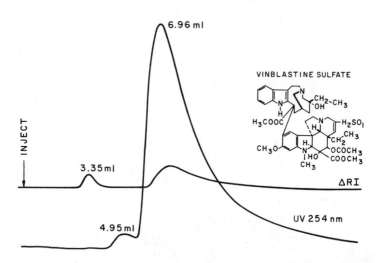

Fig. 6–57. Vinblastine. *Instrument*: Waters ALC 202/401. *Column*: 4 ft × 2.3 mm; packed with Corasil I. *Eluent*: 0.01 M sodium acetate and 0.05 M sodium sulfate. *Flow rate*: 0.37 ml/min. *Stationary phase*: duraphase-1-propylamine. [Contributed by Waters Associates (1970).]

tography separation requiring 19 hours, these compounds can be separated quantitatively in 10 minutes by high pressure liquid chromatography (Fig. 6–56). A compound that is difficult to analyze because of its high molecular weight and complicated structure is vinblastine, an antineoplastic agent. The research group at Waters Associates was able to obtain an excellent chromatogram of vinblastine using partition chromatography with an eluent of 0.01 *M* sodium acetate and 0.05 *M* sodium sulfate and a column packed with Corasil I with a stationary phase of duraphase-1-propylamine (Fig. 6–57). Many more separations of drugs and their metabolites have been obtained by high pressure liquid chromatography and others will be achieved in the near future. The examples cited are merely illustrations of the type of analyses, the kind of conditions used, and the amount of time that can be saved by utilizing this technique.

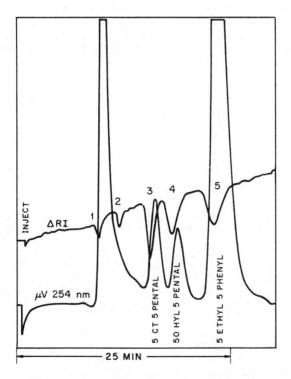

Fig. 6–58. Impurities in barbiturates. *Instrument*: Waters ALC 202/401. *Column*: 6 ft × 2.3 mm; packed with Corasil II. *Eluent*: chloroform. *Flow rate*: 0.48 ml/min. *Sample*: (1) impurity; (2) impurity; (3) pentobarbitol; (4) secobarbitol; (5) phenobarbitol. [Contributed by Waters Assoc., Inc. (1970).]

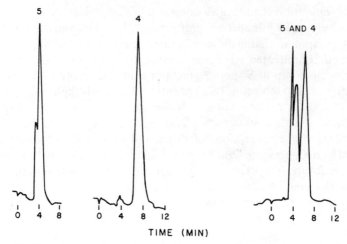

TIME (MIN)

Fig. 6–59. Dye intermediates. *Instrument*: Nester/Faust Model 1200. *Column*: 3 mm × 50 cm; packed with Sil-x. *Solvent*: 10:1, THF:water. *Flow rate*: 0.82 ml/min. *Pressure*: 200 psi. *Detector*: UV. *Attenuation*: 0.1 o.d. *Sample*: sulfanilic acid and 1-(4-sulfophenyl)-3-carboxy-5-hydroxyl-pyrazalone. [Contributed by Perkin-Elmer Corp. (1970), formerly Nester/Faust Mfg. Corp.]

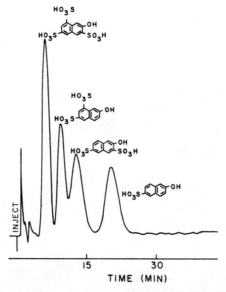

TIME (MIN)

Fig. 6–60. Dye intermediates. *Instrument*: Waters ALC 202. *Column*: 2 ft × 2.3 mm; packed with high capacity ion-exchanger (WBAX-10). *Eluent*: 0.02 N sodium citrate, pH 4.25; 0.05 M sodium sulfate, pH 4.25. *Flow rate*: 1.2 ml/min. *Temperature*: 50°C. *Samples*: mono-, di-, and trisulfonated salts of 2-hydroxynaphthalene. [Contributed by Waters Associates, Inc. (1970).]

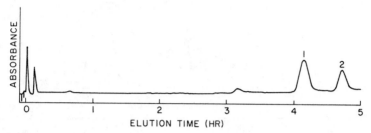

Fig. 6–61. Pear extract. *Instrument*: Waters ALC 100. *Column*: glass-50 cm × 0.9 cm; packed with 100 mesh silicic acid. *Indicator*: O-nitrophenol in absolute methanol. *Flow rate*: 3.3 ml/min. *Gradient*: $CHCl_3$ in *tert*-alcohol, 7% *tert*-alcohol in $CHCl_3$, 7% *tert*-alcohol in $CHCl_3$, 30% *tert*-alcohol in $CHCl_3$, and 50% *tert*-alcohol in $CHCl_3$. *Peaks*: (1) acetic acid; (2) malic acid; (3) citric acid. [Contributed by Waters Associates, Inc. (1970).]

D. Industrial Use of High Pressure Liquid Chromatography

In industry, high pressure liquid chromatography is used to assay drugs, check impurities in pharmaceuticals and foods determine levels of flavors and preservatives, and separate intermediates in manufacturing processes. For example, the impurities in three barbiturates are shown in Fig. 6–58. Dye intermediates, sulfanilic acid, and 1-(4-sulfophenyl)-3-carboxy-5-hydroxylpyrazalone were separated by Nester/Faust workers and are demonstrated in Fig. 6–59. Using a high-capacity ion exchanger in the column, the workers at Waters Associates separated the mono-, di-, and trisulfonated salts of 2-hydroxynaphthalene which are also dye intermediates (Fig. 6–60). The organic acids in pear extract and beer extract were obtained by the same group (Fig. 6–61 and 6–62) and were compared to standards of organic acids (Fig. 6–63).

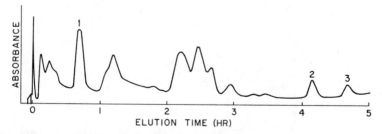

Fig. 6–62. Beer Extract. *Instrument*: Waters ALC 100. *Column*: 0.9 cm × 50 cm; packed with 100 mesh silicic acid. *Indicator*: 6-nitrophenol in absolute methanol. *Eluents*: (1) $CHCl_3$; (2) 7% *tert*-alcohol in $CHCl_3$; (3) 7% *tert*-alcohol in $CHCl_3$; (4) 30% *tert*-alcohol in $CHCl_3$; (5) 50% *tert*-alcohol in $CHCl_3$. *Flow rate*: 3.3 ml/min. *Indicator*: 0.9 ml/min. *Peaks*: (1) malic acid; (2) citric acid. [Contributed by Waters Associates, Inc. (1970).]

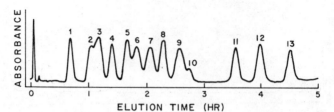

Peak	Acid	Elution volume (ml)	Peak	Acid	Elution volume (ml)
1	Acetic	127	8	Succinic	449
2	Pyruvic	205	9	α-Ketoglutaric	505
3	Formic	225	10	*Trans*-Aconitic	541
4	Fumaric	269	11	*Cis*-Aconitic	704
5	Glutaric	327	12	Malic	786
6	β-Hydroxybutyric	354	13	Citric	891
7	Lactic	404			

Fig. 6–63. Standard for organic acids.
Instrument: Waters ALC 100. *Column*: 0.9 × 50 cm; packed with 100 mesh silicic acid. *Indicator*: O-nitrophenol in absolute methanol. *Eluent*: (1) $CHCl_3$; (2) 7% *tert*-alcohol in $CHCl_3$; (3) 7% *tert*-alcohol in $CHCl_3$; (4) 30% *tert*-alcohol in $CHCl_3$; (5) 50% *tert*-alcohol in $CHCl_3$. *Flow rate*: 3.3 ml/min. [Contributed by Waters Associates, Inc. (1970).]

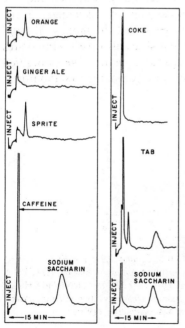

Fig. 6–64. Sodium saccharin. *Instrument*: waters ALC-202. *Column*: 4ft × 2.3 mm; packed with WBAX-10. *Eluent*: 0.1 *M* citric acid, pH 2.3. *Temperature*: 51°C. *Pressure*: 2800 psig. [Contributed by Waters Associates, Inc. (1970).]

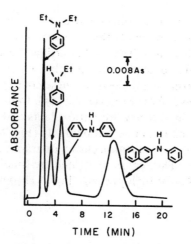

Fig. 6–65. Plasticizers. *Instrument*: noncommercial. *Column*: 1000 mm × 2.1 mm i.d.; packed with 0.5% β,β'-oxydipropionitrile on 20–37 μm Zipax support. *Eluent*: isooctane. *Flow rate*: 0.31 ml/min. *Sample*: 10.6 μl of a mixture of 9.5 μg/ml each of N,N-diethylaniline and N-ethylaniline and 29 μg/ml of diphenylamine and 52 μg/ml of N-phenyl-2-naphthylamine in isooctane. [Contributed by Majors (1970).]

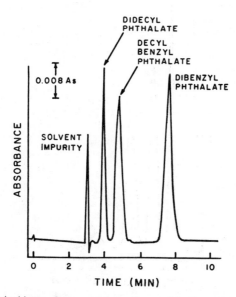

Fig. 6–66. Antioxidants. *Instrument*: noncommercial. *Separation mode*: separation of aromatic amine antioxidants using OPN/Durapak. *Column*: 1000 mm × 2.1 mm i.d.; packed with 3.7% OC₂H₄CN on 36–75 μm Porasil G; Carrier: isooctane. *Flow rate*: 2.24 ml/min. *Sample*: Aromatic amine antioxidants. [Contributed by Majors (1970).]

Because flavor strength in artificially sweetened beverages depends more on apparent sweetness and acidity than in sugar-sweetened beverages, it is important to be able to control the level of artificial sweeteners. This is possible with high pressure liquid chromatography, which provides a rapid and direct method of determining the level of these compounds, the caffeine or the preservatives in beverages or food (Fig. 6–64). Two classes of compounds that are used widely in industry are antioxidants and plasticizers. Majors (1970) separated aromatic amine antioxidants (Fig. 6–65) and phthalate plasticizers (Fig. 6–66) using partition chromatography on a noncommercial liquid chromatograph. The separations cited are simply demonstrations of the uses and versatility of high pressure liquid chromatographic systems.

E. POLLUTION CONTROL

Because of intense ecological interest in pesticide composition and residues, there was a need for a system to analyze these substances rapidly,

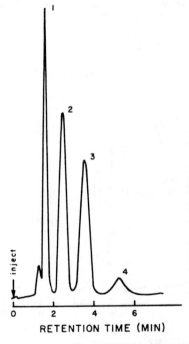

Fig. 6–67. Herbicides. *Instrument*: du Pont 820 liquid chromatograph. *Column*: 1 m × 2.1 mm i.d.; packed with β,β'-oxydipropionitrile (BOP). *Mobile phase*: dibutyl ether. *Pressure*: 300 psig. *Detector*: UV photometer at 254 nm. *Samples*: (1) linuron; (2) diuron; (3) monuron; (4) fenuron. [Contributed by du Pont Instrument Products Division, (1970).]

sensitively, and accurately. Du Pont workers obtained good separation of a group of herbicides (Fig. 6–67). The pesticides, DDT, dieldrin, methoxychlor, and 2,4,-D were separated in only a few minutes by the group at Nester/Faust (Fig. 6–68). In pollution studies it is also necessary to be able to analyze quickly and efficiently the water systems and the sewage effluent. Using an automatic organic acid analyzer, sewage effluent was analyzed for organic acids (Fig. 6–69). There are also times when it is necessary to analyze waste products. In Fig. 6–70 an automatic organic acid analyzer was used to analyze horse excrement for organic acids.

F. CHEMICAL RESEARCH

In organic, inorganic, and analytical chemistry, as well as in biochemistry, the need to analyze polar, nonvolatile compounds quickly, sensitively, and accurately has been apparent to those in the field for many years. Although gas chromatography has filled the bill to some extent through the use of derivatization, the extra steps involved tend to make it slower and more

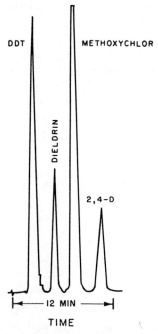

Fig. 6–68. Pesticides. *Instrument*: Nester/Faust Model 1200 liquid chromatograph. *Column*: 3 mm × 50 cm; packed with Sil-X adsorbent. *Solvent*: 5:1, hexane: chlorobutane. *Flow rate*: 1 ml/min. *Pressure*: 250 psi. *Detector*: UV. *Attenuation*: 0.1 o.d. [Contributed by Perkin-Elmer Corp. (1970), formerly Nester/Faust Mfg. Corp.]

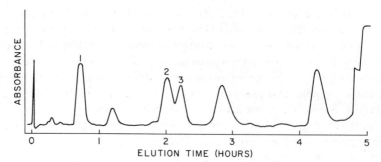

Fig. 6–69. Sewer effluent. *Instrument*: Waters ALC 100. *Column*: 0.9 × 50 cm; packed with 100 mesh silicic acid. *Indicator*: O-nitrophenol in absolute methanol. *Gradient*: (1) CH_3Cl; (2) 7% *tert*-alcohol in $CHCl_3$: (3) 7% *tert*-alcohol in $CHCl_3$; (4) 30% *tert*-alcohol in $CHCl_3$; (5) 50% *tert*-alcohol in $CHCl_3$. *Flow rate*: 3.3 ml/min. *Samples*: (1) acetic acid: (2) lactic acid; (3) succinic acid. [Contributed by Waters Associates (1970).]

cumbersome. Since the only requirement for the use of high pressure liquid chromatography is that the solutes must be soluble in some solvent, the limitations of this method are very small. The only other requirement is that the right conditions be found to achieve the best possible separation in the shortest time. In a few years, when the literature builds up, this will not be a problem; now, however, the process may be time-consuming since "trial and error" must be combined with some educated guesses as to what kind of conditions are best for the particular analytical problem at hand.

The excellent separation of six-membered carbon compounds, *n*-hexane, hexene-1, cyclohexane, and benzene was achieved by the group at Waters Associates using UV and differential refractometers as dual detectors (Fig. 6–71). Polynuclear aromatic hydrocarbons have been resolved by several

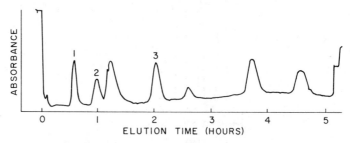

Fig. 6–70. Horse excrement. *Instrument*: Waters ALC 100. *Column*: 0.9 × 50 cm; packed with 100 mesh silicic acid. *Indicator*: O-nitrophenol in absolute methanol. *Gradient*: (1) $CHCl_3$; (2) 7% *tert*-alcohol in $CHCl_3$; (3) 7% *tert*-alcohol in $CHCl_3$; (4) 30% *tert*-alcohol in $CHCl_3$; (5) 50% *tert*-alcohol in $CHCl_3$. *Flow rate*: 3.3 ml/min. *Samples*: (1) acetic acid; (2) pyruvic acid; (3) lactic acid. [Contributed by Waters Associates, Inc. (1970).]

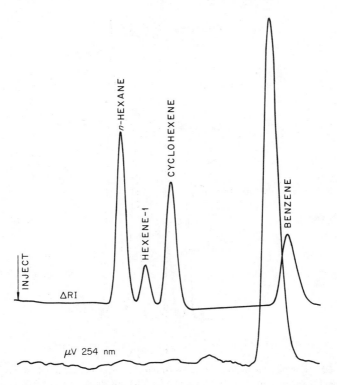

Fig. 6–71. Benzene, cyclohexane, hexene-1, *n*-hexane. *Instrument*: Waters ALC 202/401. *Column*: 4 ft × 2.3 mm; packed with Porasil T. *Eluent*: perfluoro (cyclohexane). *Flow rate*: 0.5 ml/ min. *Pressure*; 970 psig. (Contributed by Waters Assoc., Inc. (1970).)

groups and an illustration of this separation as obtained by the Nester/Faust group is shown in Fig. 6–72. A chromatogram of the analysis of chlorinated benzenes from Waters Associates is demonstrated in Fig: 6–73. The phthalic acid isomers, phthalic terepthalic and isophthalic acids, were separated by du Pont workers in 11 minutes using a strong anion exchange column and a gradient starting with distilled water; pH and ionic strength were varied to achieve excellent resolution (Fig. 6–74). The same isomers were separated by the group at Nester/Faust using adsorption chromatography (Fig. 6–75). The aliphatic alcohols, amyl alcohol, propanol, and ethanol were separated by the Varian group using a microadsorption detector (Fig. 6–76) and the group at du Pont obtained a good chromatogram of alcohols which contained a benzene ring using partition chromatography and a UV detector at 254 nm (Fig. 6–77). Aromatic amines were separated by the same group using a strong cation exchange resin and a mobile phase of sodium nitrate

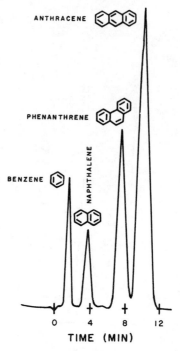

Fig. 6–72. Polynuclear hydrocarbons. *Instrument*: Nester/Faust Model 1200 liquid chromatograph. *Column*: 3 mm × 50 cm; packed with Sil-x adsorbent. *Solvent*: n-hexane (100%). *Flow rate*: 0.82 ml/min. *Pressure*: 200 psi *Detector*: UV. *Attenuation*: 0.1 o.d. *Samples*: (1) benzene; (2) naphthelene; (3) phenanthrene; (4) anthracene. [Contributed by Perkin-Elmer Corp. (1970), formerly Nester/Faust Mfg. Corp.]

(Fig. 6–78). They obtained a good chromatogram of the hydroquinones using partition chromatography (Fig. 6–79). Separation of anthraquinones is shown in Fig. 6–80. Triphenyl phosphate, tricresyl phosphate, and dibutyl phthalate were resolved and several unknown peaks were noted in the chromatogram (Fig. 6–81). The separation of carbaryl from it primary decomposition product is shown in Fig. 6–82 and the food dye, yellow 5, was separated by the Nester/Faust group from its intermediates, pyrazalone and sulfamilic acid (Fig. 6–83).

A variety of trace organic-soluble metals can be separated by partition chromatography. These metals are extracted with compounds such as dialkyldithiocarbonate which form chelates that are soluble in organic solvents. An example is shown in the Waters Associates chromatogram, Fig. 6–84. Chromium carbonyls have been separated by Burtis and Gere as is shown in Fig. 6–85. Another group of compounds, which are of chemical and bio-

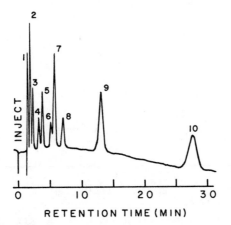

Fig. 6–73. Chlorinated benzenes. *Instrument*: du Pont 820 liquid chromatograph. *Column*: 1 m × 2.1 mm i.d.: packed with Permaphase ODS (octadecyl silane). *Mobile phase*: 50% water/ 50% methanol (v/v). *Temperature*: 60°C. *Pressure*: 1200 psig. *Flow rate*: 2 ml/min. *Detector*: UV photometer at 254 nm. *Peak identity*: (1) benzene; (2) monochlorobenzene; (3) *o*-dichlorobenzene; (4) 1,2,3-trichlorobenzene; (5) 1,3,5-trichlorobenzene; (6) 1,2,4-trichlorobenzene; (7) 1,2,3,4-tetrachlorobenzene; (8) 1,2,4,5-tetrachlorobenzene; (9) pentachlorobenzene; (10) hexachlorobenzene. [Contributed by du Pont Instrument Products Division (1970).]

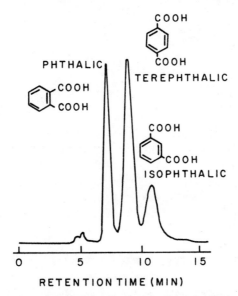

Fig. 6–74. Phthalic acid isomers. *Instrument*: du Pont 820 liquid chromatograph. *Column*: 1 m × 2.1 mm i.d.: *Temperature*: ambient. *Flow rate*: 1.5 ml/min. *Pressure*: 1400 psig. *Detector*: precision photometer at 254 nm. (Contributed by du Pont Instrument Products Division (1970).)

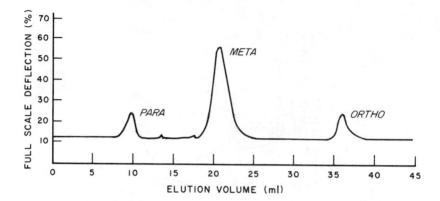

Fig. 6–75. Isomers of phthalic acid (adsorption chromatography). *Instrument*: Nester/ Faust 1200 liquid chromatograph. *Column*: 10 in. × 10 mm; packed with Corasil silica gel (325 mesh). *Eluent*: methanol. *Flow rate*: 0.6 ml/min. *Detector*: UV. *Attenuation*: 64 ×. [Contributed by Perkin-Elmer Corp. (1970), formerly Nester/Faust Mfg. Corp.]

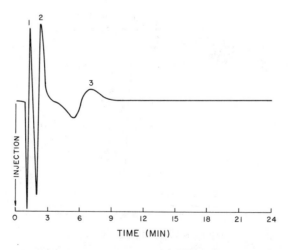

Fig. 6–76. Aliphatic alcohols. *Instrument*: Varian LCS 1000. *Column*: 6 in × ⅛ in.; packed with 74 dp 105 μm glass beads. *Sample size*, 30 μg total; (1) amyl alcohol; (2) propanol; (3) ethanol. *Attenuation*: 64 ×. *Liquid phase*: water. *Carrier*: n-heptane. *Flow*: 20 cm³/hr. *Detector*: porous glass adsorbent and glass beads. [Contributed by Varian Associates (1970).]

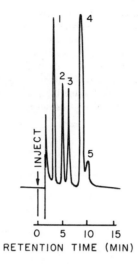

Fig. 6–77. Alcohols containing benzene ring. *Instrument*: du Pont 820 liquid chromatograph. *Column*: 1 m × 2.1 mm i.d.; packed with trimethylene glycol (TMG). *Mobile phase*: *n*-heptane saturated with TMG. *Pressure*: 600 psig. *Detector*: UV photometer at 254 nm. *Peak identity*: (1) α,α-dimethyl benzyl alcohol; (2) α-methyl benzyl alcohol; (3) 2-phenylethyl alcohol; (4) cinnamyl alcohol; (5) benzyl alcohol. [Contributed by du Pont Instruments Products Division (1970).]

chemical interest and have been separated quickly, efficiently, and quantitively, are the amide, nitrile, and carboxylic isomers of pyridine by Talley (1971). The separation of the amides and nitriles are shown in Fig. 6–86. For the amides, a nitrile was used as an internal standard and for the nitrile, nicotinanide.

These are just a few illustrations of the ways high pressure liquid chromatography can be used in chemical research. In addition to being able to detect a wide variety of metallic compounds, it may be used to detect decomposition of a material, chemical change through polymerization and chemical contamination, and to monitor steps in organic chemical syntheses.

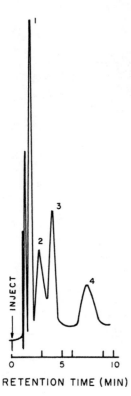

RETENTION TIME (MIN)

◀ **Fig. 6–78.** Aromatic amines. *Instrument*: du Pont 820 liquid chromatograph. *Column*: 1m × 2.1 mm i.d.; strong cation exchange (SCX). *Mobile phase*: 0.15 *M* sodium nitrate. *Pressure*: 1200 psig. *Detector*: UV photometer at 254 nm. *Peak identity*: (1) pyridine; (2) 8-hydroxyquinoline; (3) quinoline; (4) isoquinoline. [Contributed by du Pont Instrument Products Division (1970).]

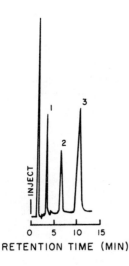

RETENTION TIME (MIN)

◀ **Fig. 6–79.** Hdroquinones. *Instrument*: du Pont 820 liquid chromatograph. *Column*: 1 m × 2.1 mm i.d.; packed with β,β'-oxydipropionitrile (BOP). *Mobile phase*: 95% heptane/5% ethanol (v/v). *Pressure*: 900 psig. *Detector*: UV photometer at 254 mm. *Peak identity*: (1) *tert*-butyl hydroquinone; (2) bromohydroquinone; (3) hydroquinone. [Contributed by du Pont Instrument Products Division (1970).]

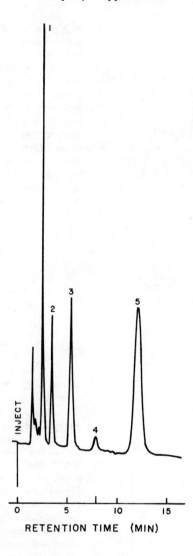

Fig. 6–80. Anthraquinones. *Instrument*: du Pont 820 liquid chromatograph. *Column*: 1 m × 2.1 mm i.d.; packed with Permaphase ODS. *Mobile phase*: 40% water/60% methanol (v/v). *Pressure*: 1200 psig. *Temperature*: 40°C. *Flow rate*: 2.2 ml/min. *Detector*: UV photometer at 254 nm. *Peak identity*: (1) 9,10-anthraquinone; (2) 2-methyl-9,10-anthraquinone; (3) 2-ethyl-9, 10-anthraquinone; (4) 1,4-dimethyl-9,10-anthraquinone; (5) 2-*tert*-butyl-9,10-anthraquinone. [Contributed by du Pont Instrument Products Division (1970).]

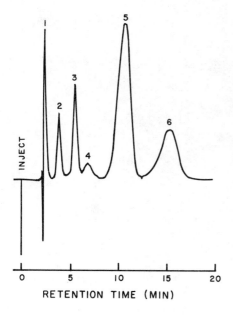

◀ **Fig. 6–81.** Phosphates and dibutyl phthalate. *Instrument*: du Pont 820 liquid chromatograph. *Column*: 1 m × 2.1 mm i.d.; packed with Permaphase ODS. *Mobile phase*: 25% isopropanol/75% water (v/v). *Pressure*: 500 psig. *Temperature*: 65°C. *Flow rate*: 1 ml/min. *Detector*: UV photometer at 254 nm. *Peak identity*: (1) solvent; (2) triphenyl phosphate; (3) dibutyl phthalate; (4) unknown; (5) tricresyl phosphate; (6) unknown. [Contributed by du Pont Instrument Products Division (1970).]

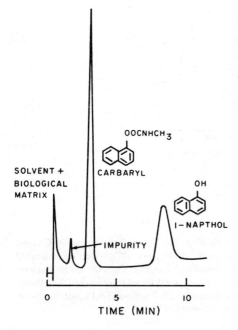

◀ **Fig. 6–82.** Plant extracts. *Instrument*: du Pont Model 820 liquid chromatograph. *Column*: 1 m × 2.1 mm; packed with Zipax chromatographic support coated with trimethylene glycol (TMG). *Precolumn*: chromosorb WAW coated with TMG. *Mobile phase*: *n*-hexane saturated with TMG. *Temperature*: ambient. *Flow rate*: 2–3 ml/min. *Pressure*: 1200 psig. *Sample*: carbaryl and 1-naphthol from a plant extract. *Sample size*: 2 μl chloroform extract. *Detector*: UV at 254 nm. *Sensitivity*: 0.04 AU full scale. [Contributed by du Pont Instrument Products Division (1970).]

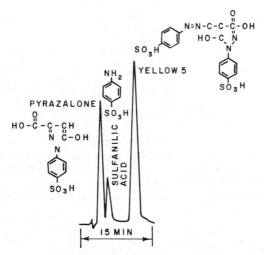

Fig. 6–83. Food dyes and intermediates: pyrazalone, sulfanilic acid, and FD + C yellow 5. *Instrument*: Nester/Faust 1200. *Column*: 3 mm × 50 cm; packed with Sil-X. *Solvent gradient*: 1:6, methanol/THF; 1.2, methanol/THF with 1% acetic acid added. *Flow rate*: 1.6 ml/min. *Pressure*: 250 psi. *Detector*: UV. *Attenuation*: 0.1 o.d. [Contributed by Perkin-Elmer Corp. (1970), formerly Nester/Faust Mfg. Corp.]

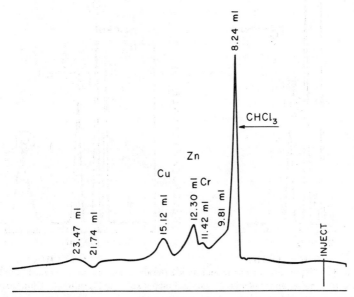

Fig. 6–84. Metal chelates. *Instrument*: Waters ALC 100. *Sample*: chelates. *Mobile phase*: cyclohexane. *Sensitivity*: 2 ×. *Flow rate*: 0.26 ml/min. *Sample load*: 10 μl. *Stationary phase*: Carbowax 400. *Column*: 6 ft × 2 mm i.d.: packed with Durapak. [Contributed by Waters Assoc. (1970).]

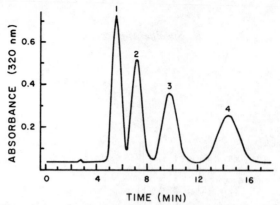

Fig. 6–85. Chromium carbonyls. *Instrument*: noncommercial. *Column*: 55 cm × 3.5 mm; packed with Carbowax 400 on PoraSil C. 36–75 μm. *Detector*: UV at 320 nm. *Eluent*: isoctane. *Flow rate*: 1.85 ml/min. *Samples*: (1) mesitylenetricarbonylchromium; (2) mexylenetricarbonyl-chromium; (3) toluenetricarbonylchromium; (4) benzenetricarbonylchromium. [Contributed by Burtis and Gere (1970).]

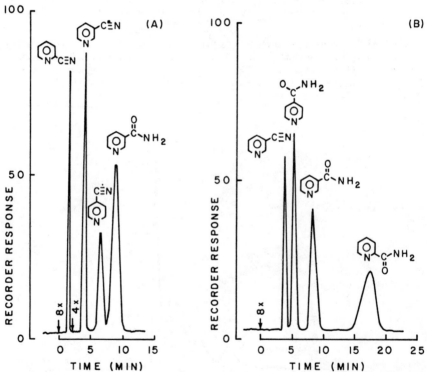

Fig. 6–86. Separation of nitrile and amide pyridine isomers. *Instrument*: du Pont Model 820 liquid chromatograph. *Column*: 2.1 mm × 1 m; packed with 1% sulfonated fluorocarbon on Zipax glass beads. *Eluent*: 0.1 *N* sodium nitrate and 0.1 *N* phosphoric acid. *Flow rate*: 1.70 ml/min. *Samples*: (A) (1) picolinonitrile: (2) nicotinonitrile; (3) isonicotinonitrile; (4) nicotin-amide (internal standard); (B) (1) nicotinonitrile (internal standard); (2) isonicotinamide; (3) nicotinamide; (4) picolinamide. [Contributed by Talley (1971).]

184

BIBLIOGRAPHY

Anders, M. W., and Latorre, J. P. (1970). *Anal. Chem.* **42**, 1430.

Anders, M. W., and Latorre, J. P. (1971). *J. Chromatogr.* **55**, 409.

Anderson, N. G. (1962). *Anal. Biochem.* **4**, 269.

Anderson, N. G., and Ladd, C. C. (1962). *Biochim. Biophys. Acta* **55**, 275.

Anderson, N. G., Green, J. G., Barber, M. L., and Ladd, F. C., Sr. (1963). *Anal. Biochem,* **6**, 153.

Anderson, R. E. (1972). *Chromatographia*, **5**, 105.

Bartlett, G. R. (1970). *Advan. Exp. Med. Biol.* **6**, 245–268.

Bergmeyer, H. H., ed. (1963). "Methods of Enzymatic Analysis," p. 543 Academic Press, New York.

Bertles, J. F., and Beck, W. S. (1962) *J. Biol. Chem.* **237**, 3770.

Bianco, T., Guistina, G., and Lazzarrini, E. (1962). *Nature, (London)* **194**, 289.

Bishop, C. (1960). *J. Biol. Chem.* **235**, 3228.

Blattner, F. R., and Erickson, H. P. (1967). *Anal. Biochem.* **18**, 220.

Bombaugh, K. J., and Levangie, R. F. (1970). *Separa. Sci.* **5**, 751.

Bombaugh, K. J., and Little, J. N. (1964). *J. Chromatogr. Sci.* **16**, 47.

Bombaugh, K. J., Dark W. A., and Levangie, R. F. (1969). In "Advances in Chromatography" (A. Zlatkis, ed.), p. 334. Preston Tech. Abstr. Evanston, Ill.

Bombaugh, K. J., Levangie, R. F., King, R. M., and Abrahams, L. (1970). *J. Chromatogr. Sci.* **8**, 657.

Brooker, G. (1970). *Anal. Chem.* **42**, 1108.

Brooker, G. (1971). *Anal. Chem.* **43**, 1095.

Brown, P. R. (1970). *J. Chromatogr.* **52**, 257.

Brown, P. R. (1971a). *J. Chromatogr.* **57**, 383.

Brown, P. R. (1971b). *Anal. Biochem.* **43**, 305.

Brown, P. R., Agarwal, R. P., Gell, J., and Parks, Jr., R. E. (1972), *Comp. Biochem. and Physiol.* (in press).

Brown, P. R., and Miech, R. P. (1972) *Anal. Chem.* **44**, 1072.

Brown, P. R., and Parks, R. E., Jr. (1971). *Pharmacologist* **13**, 210.

Burtis, C. A. (1970a). *J. Chromatogr.* **52**, 97.

Burtis, C. A. (1970b). *J. Chromatogr.* **57**, 183.

Burtis, C. A., and Gere, D. (1970). "Nucleic Acid Constituents by Liquid Chromatography." Varian Aerograph, California.

Burtis, C. A., and Warren, K. S. (1968). *Clin. Chem.* **14**, 290.

Burtis, C. A., Butts, W., Rainey, W., and Scott, C. D. (1970a). *Amer. J. Clin. Pathol.* **53**, 769.

Burtis, C. A., Gere, D., and MacDonald, F. (1970b). *Chromatographia* **3**, 116.

Burtis, C. A., Munk, M. N., and MacDonald, F. R. (1970c). *Clin. Chem.* **16**, 667.

Busch, E. W. (1968). *J Chromatogr.* **37**, 518.

Byrne, S. H., and Schmit, J. A. (1971). *J. Chromatograph. Sci.* **9**, 592.

Caldwell, I. (1969). *J. Chromatogr.* **44**, 331.

Chalfin, D. (1956). *J. Cell. Comp. Physiol.* **47**, 215.

Chilcote, D. D., and Mrocher, I. E. (1971). *Clin. Chem.* **17**, 751.

Clifford, A. J., Riumallo, J. A., Brown, P. R., and Scrimshaw, N. S. (unpublished data).

Clifford, A. J., Riumallo, J. A., Baliga, B. S., Munro, H. N., and Brown, P. R. (1972). *Biochim. Biophys. Acta* **277**, 443.

Cohn, W. E. (1949). *Science* **109**, 377.

Cohn, W. E. (1950). *J. Amer. Chem. Soc.* **72**, 1471.

Cohn, W. E. (1955). *In* "Nucleic Acids, Chemistry and Biology" (E. Chargaff and J. N. Davidson, eds.), Vol. I, p. 212. Academic Press, New York.

Cohn, W. E., and Bollum, F. J. (1961). *Biochem. Biophys. Acta* **48**, 588.

Conlon, R. D. (1969). *Anal. Chem.* **41**, 107A.

Crampton, C. F., Frankel, F. R., Benson, A. M., and Wade, A. (1960). *Anal. Biochem.* **1**, 249.

Davenport, T. B. (1969). *J. Chromatog.* **42**, 219.

DeStefano, J. J., and Beachell, H. C. (1970). *J. Chromatogr. Sci.* **8**, 434.

Dixon, M., and Webb, E., eds. (1964). *In* "The Enzymes" 2nd ed., p. 716. Academic Press, New York.

Dorfman, L. (1953). *Chem. Rev.* **53**, 47.

Duch, D., and Laskowski, K. (1971). *Anal. Biochem.* **44**, 42.

Dybeznski, R. (1967). *J. Chromatogr.* **31**, 155.

Felton, H. (1969). *J. Chromatogr. Sci.* **7**, 13.

Gehrke, C. W., and Ruyle, C. D. (1968). *J. Chromatogr.* **38**, 473.

Gere, D. (1970). Unpublished data, courtesy Varian Aerograph.

Giddings, J. C. (1965). "Dynamics of Chromatography," p. 190. Marcel Dekker, New York.

Gill, J. M. (1969). *J. Chromatogr. Sci.* **7**, 731.

Gordon, S. M., Krige, G. J., Haaroff, P. C., and Pretorius, V. (1963). *Anal. Chem.* **35**, 1537.

Green, J., Nunley, C., and Anderson, N. G. (1966). *Nat. Cancer Inst., Monogr.* **21**, 431.

Hadden, N., Baymann, F., MacDonald, F., Munk, M., Stevenson, R., Gere, D., and Zamaroni, F. (1971). "Basic Liquid Chromatography." Varian Aerograph, California.

Halász, I., and Horvath, C. (1964a). *Anal. Chem.* **36**, 1178.

Halász, I., and Horvath, C. (1964b). *Anal. Chem.* **36**, 2228.

Halász, I., and Naefe, M. (1972). *Anal. Chem.* **44**, 76.

Halász, I., and Walking, P. (1969), in "Advances in Chromatography" (A. Zlatkis, ed.), p. 310. Preston Tech. Abstr., Evanston, Illinois.

Halász, I., Gerlach, H. O., Kroneisen, A., and Walking, P. (1968). *Anal. Chem.* **234**, 82.

Hamilton, P. B. (1962). *Ann. N. Y. Acad. Sci.* **102**, 55.

Hamilton, P. B. (1963). *Anal. Chem.* **35**, 2055.

Hamilton, P. B. (1965). *Fed. Proc., Fed. Amer. Soc. Exp. Biol.* **24**, 656.

Hegedrus, L. L., and Petersen, E. E. (1971). *J. Chromatogr. Sci.* **9**, 551.

Henry, R. A., and Schmit, J. A. (1970). *Chromatographia* **3**, 116.

Henry, R. A., Schmit, J. A., Dieckman, J. F., and Murphey, J. F., (1971a). *Anal. Chem.* **43**, 1053.

Henry, R. A., Schmit, J. A., and Dieckman, J. F. (1971b). *J. Chromatogr. Sci.* **9**, 513.
Hesse, G., and Englehardt, H. (1966). *J. Chromatogr.* **21**, 228.
Hillman, D. E. (1971). *Anal. Chem.* **43**, 1007.
Horgan, Jr., D. F., and Little, J. N. (1972). *J. Chromatographic Sci.*, **10**, 76.
Hornig, M. J. (1968). "Biomedical Applications of Gas Chromatography," Vol. II, pp. 53–86. Plenum, New York.
Horvath, C. G. (1972a). "Methods of Biochemical Analysis" (in press).
Horvath, C. G. (1972b). *In* "Ion Exchange," Vol. 3. Dekker, New York (in press).
Horvath, C. G., and Lipsky, S. R. (1966). *Nature,* (*London*) **211**, 748.
Horvath, C. G., and Lipsky, S. R. (1969a). *In* "Advances in Chromatography" (A. Zlatkis, ed.). p. 310. Preston Tech. Abstr., Evanston, Illinois.
Horvath, C. G., and Lipsky, S. R. (1969b). *J. Chromatogr. Sci.* **7**, 109.
Horvath, C. G., and Lipsky, S. R. (1969c). *Anal. Chem.* **41**, 1227.
Horvath, C. G., Preiss, B. A., and Lipsky, S. R. (1967). *Anal. Chem.* **39**, 1422.
Howard, C., and Martin, A. (1950). *Biochem. J.* **56**, 532.
Huber, J. F. K. (1969a). *In* "Advances in Chromatography" (A. Zlatkis, ed.), p. 283. Preston Tech. Abstr., Evanston, Illinois.
Huber, J. F. K. (1969b). *J. Chromatogr. Sci.* **7**, 85.
Huber, J. F. K. (1971). *J. Chromatogr. Sci.* **9**, 72.
Huber, J. F. K., and Hulsman, J. (1967). *Anal Chem. Acta* **38**, 305.
Huber, J. F. K., Meyers. C. A. M., and Hulsman, J. (1972). *Anal. Chem.* **44**, 111.
Hupe, K. P., and Bayer, E. S. (1967). *J. Gas Chromatogr.* **5**, 197.
Hutchinson, W. C., and Munro, H. N. (1961). *Analyst* **86**, 768.
Jacobson, M., O'Brien, J. F., and Hedgcoth, C. (1968). *Anal. Biochem.* **25**, 363.
Jaffee, J. (1971). *Nature (London),* **230**, 408.
James, A. T. Ravenhill, D. R., and Scott, R. P. W. (1964). *Chem. Ind.* **18**, 146.
Jentoff, R. E., and Gouw, T. W. (1968). *Anal. Chem.* **40**, 923.
Jolley, R. L., and Scott, C. D. (1970a). *Clin. Chem.* **16**, 687.
Jolley, R. L., and Scott, C. D. (1970b). *J. Chromatogr.* **47**, 272.
Joynes, P. L., and Maggs, R. J. (1970). *J. Chromatogr. Sci.* **8**, 427.
Junowics, E., and Spenser, J. (1968). *J. Chromatogr.* **37**, 518.
Kaizuma, H., Myers, M. N., and Giddings, J. C. (1970). *J. Chromatogr. Sci.* **8**, 630.
Karger, B., and Berry, L. V. (1972). *Anal. Chem.* **44**, 93.
Karmen, A., Kane, L. D., Karasek, M., and Lapidus, B. (1970). *J. Chromatogr. Sci.* **8**, 439.
Katz, S., and Burtis, C. A. (1969). *J. Chromatogr.* **40**, 270.
Katz, S., and Comb, D. G. (1963). *J. Biol. Chem.* **328**, 3065.
Kelley, W. N., and Wyngaarden, J. (1970), *Clin. Chem.* **16**, 707.
Kemula, W. (1952). *Rocz. Chem.* **26**, 281.
Kirkland, J. J. (1968). *Anal. Chem.* **40**, 391.
Kirkland, J. J. (1969a). *Anal. Chem.* **41**, 218.
Kirkland, J. J. (1969b). *In* "Advances in Chromatography" (A. Zlatkis, ed.), p. 328. Preston Tech. Abstr., Evanston, Illinois.
Kirkland, J. J. (1970). *J. Chromatogr. Sci.* **8**, 72.
Kirkland, J. J., ed., (1971a). "Modern Practice of Liquid Chromatography." Wiley (Interscience), New York.
Kirkland, J. J. (1971b). *Anal. Chem.* 36A.
Kirkland, J. J. (1971c). *J. Chromatogr. Sci.* **9**, 206.
Kirkland, J. J. (1972). *J. Chromatogr. Sci.*, **10**, 129.
Kirkland, J. J., and DeStefano, J. J. (1969). *In* "Advances in Chromatography" (A. Zlatkis, ed.), p. 397. Preston Tech. Abstr., Evanston, Illinois.

Kirkland, J. J., and DeStefano, J. J. (1970). *J. Chromatogr. Sci.* **8**, 309.

Kirkland, J. J., and Felton, H. (1969). *J. Chromatogr. Sci.* **7**, 7.

Knox, J. H. (1966). *Anal. Chem.* **38**, 253.

Knox, J. H., and Parcher, J. F. (1969). *Anal. Chem.* **41**, 1599.

Koen, J. G., Huber, J. F. K., Poppe, H., and den Boef, G. (1970). *J. Chromatogr. Sci.* **8**, 192.

Kuby, S., Noda, T., and Lardy, H. (1953). *J. Biol. Chem.* 21 , 202.

Lange, H. W. (1970). *Anal. Biochem.* **38**, 94.

Leach, A. A., and O'Shea, P. C. (1965). *J. Chromatogr.* **17**, 245.

Leitch, R. E. (1971). *J. Chromatogr. Sci.* **9**, 531.

Le Rosen, A. L., and Rivet, C. A. (1948). *Anal. Chem.* **20**, 1093.

Lie Tien, (1953). *Anal. Chem.* **25**, 1235.

Little, J. N., Waters, J. L., Bombaugh, K. J., and Paulis, W. J. (1969). *J. Polym. Sci., Part A-2* **7**, 1775.

Little, J. N., Horgan, D. F., and Bombaugh, K. J. (1970a). *J. Chromatogr. Sci.* **8**, 625.

Little, J. N., Waters, J. L., Bombaugh, K. J., and Paulis, W. J. (1970b). *Separ. Sci.* **5**, 765.

Little, J. N., Waters. J. L., Bombaugh, K. J., and Paulis, W. J. (1971). *J. Chromatogr. Sci.* **9**, 341.

Locke, D. C. (1967). *Gas Chromatogr.* **5**, 202.

Locke, D. C. (1968). *J. Chromatogr.* **35**, 24.

Locke, D. C. (1969). *In* "Advances in Chromatography" (A. Zlatkis, ed.), p. 47. Preston Tech. Abstr. Evanston, Illinois.

Locke, D. C., Schmermund, J. T., and Banner, B. (1972). *Anal. Chem.* **44**, 90.

Loring H. S. (1955). *In* "The Nucleic Acids" (E. Chargaff and J. N. Davidson, eds.), Vol. 1, Academic Press, New York.

Lowy, B., Williams, M. K., and London, I. M. (1962). *J. Biol. Chem.* **237**, 1622.

Lyons, J. G. (1972). *Chromatographia* **5**, 156.

McNair, H. M., and Bonelli, E. J. (1968). "Basic Gas Chromatography." Varian Aerograph, California.

Maggs, R. J. (1968). *Chromatographia* **1**, 43.

Maggs, R. J. (1969). *In* "Advances in Chromatography" (A. Zlatkis, ed.), p. 303. Preston Tech. Abstr., Evanston, Illinois.

Majors, R. E. (1969). *In* "Advances in Chromatography" (A. Zlatkis, ed.) p. 406. Preston Tech. Abstr., Evanston, Illinois.

Majors, R. E. (1970). *J. Chromatogr. Sci.* **8**, 338.

Mandel, P. (1964) *Progr. Nucl. Acid Res. Mol. Biol.* **8**, 304.

Manley, F., and Manley, G. (1960). *J. Biol. Chem.* **235**, 235.

Martin, A., and Synge, R. (1941). *Biochem. J.* **35**, 1358.

Martire, D. E., and Locke, D. C. (1971). *Anal. Chem.* **43**, 68.

Moore, J. C. (1964). *J. Polym. Sci., Part A* **2**, 835.

Moore, S., and Stein, W. (1957). *J. Biol. Chem.* **192**, 663.

Mrochek, J. E., Butts, W. C., Rainey, W. T., Jr., and Burtis, C. A. (1971). *Clin. Chem.* **17**, 72.

Muller, R. (1968). *Anal. Chem.* **40**, 109A.

Munch, P. A., and Kalchai, H. (1963). *In* "Methods in Enzymology" (S. P. Colowick and N. O. Kaplan, eds.), Vol. 6, p. 869. Academic Press. New York.

Munk, M. N., and Ravel, D. N. (1969). *J. Chromatograph. Sci.* **7**, 48.

Munro, H. N., and Fleck, A. (1964). *Methods Biochem. Anal.* **17**, 118.

Murokani, F. (1971). *J. Chromatogr.* **53**, 584.

Nelson, J. A., and Parks, Jr., R. E. (1972). *Cancer Res.* (in press).

Noda, T., and Kuby, S. (1963). *In* "Methods in Enzymology" Vol. 6, 223. Academic Press, New York.

Obrink, B., Lorvasd, L. L., and Rigley, R. (1967). *J. Chromatogr.* **31**, 48.

Pennington, S. (1971). *Anal. Chem.* **43**, 1701.

Pitts, W. W., Jr., Scott, C. D., Johnson, W. F., and Jones, G., Jr. (1970). *Clin. Chem.* **16**, 657.

Ramsey, R.; and Warren, C. O., Sr. (1933). *Quart. J. Exp. Physiol.* **22**, 49.

Randerath, E., Ten Broeke, J. W., and Randerath, K. (1968). *FEBS Lett.* **2**, 10.

Randerath, K. (1963). "Thin-Layer Chromatography," 1st ed., p. 189. Academic Press, New York.

Randerath, K., and Randerath, E. (1967). In "Methods in Enzymology" (L. Grossman and K. Moldave, eds.), Vol. 12, Part A, p. 323. Academic Press, New York.

Reefrey, V. G., and Mirsky, A. E. (1952). *J. Gen. Physiol.* **35**, 841.

Robinson, G., Butcher, R., and Sutherland, E. (1968). *Ann. Amer. Rev. Biochem.* **37**, 149.

Rogers, L. B. (1959). *In* "Treatise on Analytical Chemistry" (P. M. Kolthoff and P. J. Elving, eds.), Part I, p. 919. (Interscience), New York.

Rony, P. R. (1968). *Separ. Sci.* (1900). **3**, 425.

Scholar, E. M., Brown, P. R., and Parks, Jr., R. E. (1972). *Cancer Res.* **32**, 259.

Scholar, E. M., Brown, P. R., Parks, Jr., R. E., and Calabresi, P., (1972a). *Blood* (in press).

Scott, C. D. (1968a). *Clin. Chem.* **14**, 14.

Scott, C. D. (1968b). *Clin. Chem.* **14**, 521.

Scott, C. D., and Lee, N. E. (1968). *J. Chromatogr.* **42**, 263.

Scott, C. D., Attril, J., and Anderson, N. G. (1967). *Proc. Soc. Exp. Biol. Med.* **125**, 181.

Scott, C. D., Chilcote, D. D., and Lee, N. E. (1972). *Anal. Chem.* **44**, 85.

Scott, C. G., and Bommer, P. J. (1969). *In* "Advances in Chromatography" (A. Zlatkis, ed.), p. 416. Preston Tech. Abstr., Evanston, Illinois.

Scott, C. G., and Bommer, P. J. (1970). *J. Chromatogr. Sci.* **8**, 446.

Scott, R. P. W. (1971a). *J. Chromatogr. Sci.* **9**, 385.

Scott, R. P. W. (1971b). *J. Chromatogr. Sci.* **9**, 449.

Scott, R. P. W., and Lawrence, J. G. (1969). *In* "Advances in Chromatography" (A. Zlatkis, ed), p. 276. Preston Tech. Abstr., Evanston, Illinois.

Scott, R. P. W., and Lawrence, J. G. (1970). *J. Chromatogr. Sci.* **8**, 446.

Scott, R. P. W., Blackburn, D. W. J., and Wilkins, T. (1967). *J. Gas Chromatogr.* **5**, 183.

Senft, A., Miech, R. P., Brown, P. R., and Senft, D. (1972). *Int. Parasitol.* **2**, 249.

Shmukler, H. (1970a). *J. Chromatogr. Sci.* **8**, 581.

Shmukler, H. (1970b). *J. Chromatogr. Sci.* **8**, 653.

Shmukler, H. (1972a). *J. Chromatogr. Sci.* **10**, 38.

Shmukler, H. (1972b). *J. Chromatogr. Sci.* **10**, 137.

Sie, S. T., and van den Hoed, N. (1969). *In* "Advances in Chromatography" (A. Zlatkis, ed.), p. 318. Preston Techn. Abstr., Evanston, Illinois.

Siggia, S., and Dishman, R. A. (1970). *Anal. Chem.* **42**, 1223.

Simon, M., Chang, H. X., and Laskowski, M., Sr. (1971). *Biochim. Biophys. Acta.* **232**, 46.

Smuts, T. W., and Pretorious, V. (1972). *Anal. Chem.* **44**, 121.

Smuts, T. W., van Niekiek, F. A., and Pretorious, V. (1967). *J. Gas Chromatog.* **5**, 190.

Smuts, T. W., van Niekiek, F. A., and Pretorious, V. (1969). *J. Chromatogr. Sci.*, **7**, 127.

Snyder, L. R. (1965). *Chromatographic Rev.* **7**, 1.

Snyder, L. R. (1967). *Anal. Chem.* **7**, 105.

Snyder, L. R. (1969a). *J. Chromatogr. Sci.* **7**, 352.

Snyder, L. R. (1969b). *In* "Advances in Chromatography" (A. Zlatkis, ed.), p. 355. Preston Tech. Abstr., Evanston, Illinois.

Snyder, L. R., and Saunders, D. L. (1969). *In* "Advances in Chromatography" (A. Zlatkis, ed.), p. 289. Preston Tech. Abstr., Evanston, Illinois.

Snyder, L. R. (1972). *J. Chromatographic Sci.* **10**, 200.

Sober, H. A., and Peterson, E. A. (1957). *In* "Exchange in Organic and Biochemistry" (C. Calmon and T. R. E. Kressman, eds.), p. 338. Wiley (*Interscience*), New York.

Stahl, K. W., Schafer, G., and Lamprecht, W. (1972). *J. Chromatogr. Sci.* **10**, 95.

Stegman, R. J., Senft, A. W., Brown, P. R., and Parks, Jr., R. E. (1972). *Biochem. Pharmacol.*, (in press).

Stevenson, R. (1971). *J. Chromatogr. Sci.* **9**, 257.

Stevenson, R. L. (1971). *Clin. Chem.*, **17**, 774.

Talley, C. P. (1971). *Anal. Chem.* **43**, 1512.

Thacker, L. H., Scott, C. D., and Pitt, W., Jr. (1970). *J. Chromatogr.* **57**, 175.

Uziel, M., and Koh, C. K. (1971). *J. Chromatogr.* **59**, 188.

Uziel, M., Koh, C. K., and Cohn, W. E. (1968). *Anal. Biochem.* **25**, 77.

Vavick, J. M., and Howell, R. (1970). *Clin. Chem.* **16**, 702.

Volkin, E., Kitym, J. X., and Cohn, W. B. (1951). *JACS* **73**, 1533.

Volkin, E., and Cohn, W. E. (1954). *Methods Biochem. Anal.* **1**, p. 287.

Volkin, E., Kyman, D. Y., and Cohn, W. E. (1951). *J. Amer. Chem. Soc.* **73**, 1533.

Walton, H. F. (1968). *Anal. Chem.* **40**, 51R.

Waters Associates. (1970). *Anal. Chem.* **42**, 18A.

Waters, J. L., Little, J. N., and Horgan, D. F. (1969). *J. Chromatogr. Sci.* **7**, 293.

Wyatt, G. R. (1955). *In* "The Nucleic Acids" (E. Chargaff and J. N. Davidson, eds.), Vol. 1, p. 243. Academic Press, New York.

Woods, R. A., and Lantz, C. D. (1970). Barber-Coleman Co.

Young, D. S. (1970). *Clin. Chem.* **16**, 681.

Author Index

Numbers in italics indicate the page on which the complete reference is listed.

Subject Index

A

Accuracy, definition, 126
Acetaminophen, *see* Decongestants
Acetic acid, *see* Organic acids
Adenine, *see* Bases, purine
Adenosine, *see* Nucleosides
Adsorbants
    porous, 45–46
    solid core porous, 45–46
Adsorption chromatography, *see* Liquid–
    solid chromatography
Alcohols
    amyl, 178
    benzyl, 179
    cinnamyl, 179
    propanol, 178
    separation of, 175, 178, 179
Aldrin, *see* Insecticides
Amines, aromatic, separation of, 4, 175,
    180
Amino acid analyzer, 152
Amino acids
    aspartic acid, 157
    glycine, 36
    phenylalanine, 36, 157
    proline, 37, 157
    separation of, 36, 152, 157
    valine, 157
O-aminoaniline, 4
Amobarbital, *see* Barbiturates
Amplification, 115
Amyl alcohol, *see* Alcohols
Analgesics
    aspirin, 77, 162, 163
    caffeine, 162, 163

    phenacetin, 77, 163
    salicylamide, 77, 163
    salicylic acid, 77
    separation of, 77, 159, 162, 163
Andosterone, *see* Steroids
Aniline, 4
Anthraquinones
    anthraquinone, 58
    2-ethylanthraquinone, 58
    2-methyl anthraquinone, 58
    2-tert-butyl anthraquinone, 58
    separation of, 58, 176, 181
Antioxidants, 171, 172
    dibenzyl phthalate, 171
    decyl benzyl phthalate, 171
    didecyl phthalate, 171
    separation of, 171, 172
Apparatus, 9
Applications, general, 128–130
Aromatic amines
    pyridine, 180
    quinolines, 180
Aspartic acid, *see* Amino acids
Aspirin, *see* Analgesics
Automation, 50, 81–82

B

Barbiturates
    amobarbital, 159
    hydroxyamobarbital, 159
    hydroxyphenobarbital, 159
    impurities in, 167, 169
    ketohexobarbital, 159
    metharbital, 160
    pentobarbital, 167